I.T.
RISK
MANAGEMENT

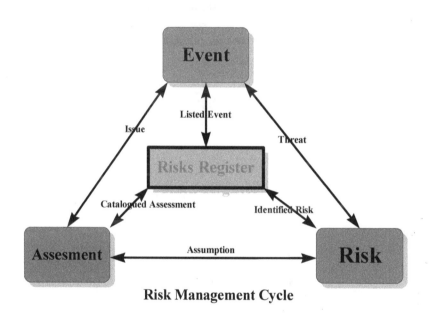

Risk Management Cycle

EurIng Prof Dr Andreas Sofroniou

Αφιερωμενο

στην *Μαμμα* μου

για το παραδειγμα της

εις το τι σημαινει να εισαι

γονιος και ρομιος.

Κικκος,

2009

... to the *mother*

who gave the good example

on how to be a parent ...

2009

Copyrights.

ISBN: 978-1-4092-7488-9

Contents

Description		Page

1. Risk Management: The Concept 11
I1 Programme Management 11
I2 The Sigma Methodology 11
I3 The Risk Management Cycle 12
I4 The Integration Of Methodologies 13
I5 The Inter-relationship 13
I6 Managing the Risk Programme 15
I7 Impact On Business 15

2. The Obligations Of A Corporation 16
I1 Inter-dependency On I.T. 16
I2 Internal Control 16
I3 Staff Training 17

3. Risk Management Analysis 18
I1 Assessment Of Risks 18
I2 Communication 18
I3 Risks Management Practice 18
I4 The Full Risk management Cycle 18

4. Safeguarding Information Systems 20
I1 Correct Software Control And Maintenance 20
I2 Access Control And Maintenance 20
I3 Firewalls 20
I4 Vulnerability Assessment 21
I5 Intrusion Detection Systems (IDS) 21
I6 Cryptography 21
I7 Virus Protection 22
I8 Content Security 22

5. Business Related Overview 23

I.T. Risk Management

I1 Business Responsibilities   23

I2 Corporate Governance   23

I3 Business Continuity   23

I4 Reputation Management   23

I5 Good Corporate Reputation   24

I6 Preparing For A Crisis   24

I7 Information Assurance   24

I8 Information Availability   25

I9 Interdependency And The Critical Information Infrastructure   25

I10 Internet Commerce   25

I11 World Wide Web (www)   26

I12 Business Case   26

I13 Global Networked Business   26

I14 Implementation of a Risk Management Strategy   27

I15 Monitoring   27

6. The Sigma Methodology Explained   28

6.1 The Programme Objectives   28

6.2 The Sigma Methodology   28

6.3 The Risk Management Cycle   30

I1 Features And Benefits Of The Sigma Approach   30

    Communication   30

    Control   30

    Information   30

    Flexible   30

    Acceptable   31

6.5 Assessment Analysis   31

6.6 Strategic Cost Analysis   31

I1 Risk Administration System Tool   32

I2 Work Plan Analysis   32

6.9 Communicating The Risks   32

6.10 How To Use The Contents Of This Book   32

I.T. Risk Management

7. The Principles Of Risk Management   33

 I1 Team Approach   33

7.2 The Definition of a Risk   33

 I2 Approaches to Risk Management   34

7.4 Events And Risk Registers   34

 I1 Individual Interviews   34

 I2 Group Brainstorming   35

 I3 Risk Analysis And Quantification of Risks   35

 I4 Risk Control and Lack of Follow-Through   35

 I5 Risk Transfer   36

7.10 Risk Management vs. Project Management   36

8. Basic Principles of the Sigma Methodology   37

 I1 The Process   37

 I2 Communication Of Assessments   37

 I3 The Current Project Plan Is The Baseline   38

8.4 Uncertainty Equals Risk   38

8.5 Judging the Quality Using the Sigma Scale   38

 I1 Sigma Process Overview   39

 I2 Project Prioritisation   40

8.8 Risk Prioritisation   40

8.9 Risk Control   41

9. Initiating the Sigma Process   43

 I1 Organisational Culture   43

 I2 Organisational Structure and Stability   43

9.3 Effect Of Current Project Status   44

9.4 Project Fully Planned And Proceeding   44

9.5 Project In Trouble   44

9.6 Interviews With Key People   44

 I1 Choosing A Suitable Risk Assessment Team   45

9.8 Risk Review Meetings   46

I.T. Risk Management

10. Project Prioritisation		47
10.1 The Project And/Or Programme	47	
10.2 Project Inclusion In Process		47
10.3 Prioritisation		47
10.4 The Project Prioritisation Process		48
10.4.1 The Standard Reference Matrix		48
10.4.2 Positioning and Prioritising New Projects		48
10.4.3 Project Approval and Resources		48
10.4.4 When Projects Change Position on the CC Diagram		49
10.4.5 Deciding which Projects Require Formal Risk Management	49	
10.5 Risk Identification and Analysis	49	
1 Assessment Analysis		50
10.5.2 Identify the Sources of Assessments		50
10.5.3 Prioritise Assessments		50
10.5.4 Convert Key Assessments Into Risks		51
1 Categorisation Of Assessments — Risk Drivers	52	
10.6 Strategic Cost Analysis		53
10.7 Work Plan Analysis		53
10.7.1 Quality Review of Plans		54
10.7.2 Locate Hot Spots in the Plans		54
10.7.3 Project Readiness Walkthroughs		54
10.7.3.1 The basic objectives of a Project Readiness Walkthrough		54
1 The process follows a standard structure.		55
10.7.3.3 The project team meets and fills in their confidence levels (A-C)		55
10.7.3.4 The Walkthrough is scheduled		55
10.7.3.5 The typical participants		55
1 The Walkthrough starts by concentrating on the Cs	55	
2 The Walkthrough is completed	56	
10.7.3.8 The matrix is updated		56
10.7.3.9 The risks reviewed regularly	56	
10.7.4 Pre-selected Assessments		56
I1 Risk Prioritisation		56

10.8.1 Assessments and Risks Register   56

10.8.2 Positioning Risks   56

10.8.3 Risk Register Reports   58

10.9 Risk Control   59

10.9.1 Strategic Approaches   59

10.9.2 Tactical Approaches   61

10.9.3 Following Through on Actions   61

1   The Need for Risk Plans   61

10.9.5 Essential Components of a Formal Risk Plan   62

10.9.6 Devising Risk Plans to Attack Risks   62

10.9.6.1 Developing Risk Plans   62

10.9.6.2 Plans for Attacking Assessment Based Risks   62

10.9.6.3 Plans for Attacking Planning-Type Risks   63

10.9.6.4 Selecting Particular Risk Plans   63

10.9.7 Project Risk Organisation And Responsibilities   65

10.9.7.1 Risk Ownership and Risk Action Managers   65

10.9.7.2 Selecting Risk Action Managers   65

10.9.7.3 Selecting Risk Owners   65

1   The Risk Review Board   66

10.9.7.5 Running Risk Plans   67

10.9.7.6 Closing Risk Plans   68

10.9.7.7 Stopping a Risk Plan   68

10.9.7.8 Closing a Risk Plan   68

10.9.7.9 Configuration Control   68

10.10   Applying The Sigma Process - Practical Considerations   69

I.T. Risk Management

10.10.1 Project Prioritisation Considerations	69	
10.10.1.1 Using software tools to maintain the project inventory		69
10.10.1.2 Assessment Analysis Considerations		70
10.10.1.3 Getting To The Right Assessment		70
10.10.1.4 Preparation For The Interview		72
10.10.1.5 During The Interview		72
10.10.1.6 At The End Of The Interview		72
10.10.1.7 Review Assessments And Risks Regularly		73
11. Transferring Ownership Of The Sigma Process		74
11.1 Sponsorship And Ownership		74
11.1.1 Identifying The Process Champions		74
11.1.2 Identifying The Process Owners		74
I1II Affording Time For The Process		74
I1III Enthusiasm And Attention To Detail	75	
I1IIII Access To Key Players In The Organisation	75	
I1IIV Avoiding The Split-up Approach		75
I2 Hand Over Process		75
I2I Training	75	
I2II Parallel Interviews		76
I2III Consultant Leading Interviews	76	
11.2.2.2 Client Leading Interviews		76
11.2.3 Procedures And Supporting Documentation		77
11.2.4 Software Tool Support		77
11.3 Ongoing Quality Management		77
11.3.1 Can they Continue To Do It Themselves?		77
11.3.2 Applying Sigma To The Hand Over Phase		78
11.4 Integrating Methodologies		78
12. Interfacing		79
12.1 The Diagrammatic Representation Of The Desired Integration	79	
I1 Suggested Interfacing		79
I2 Using The Existing Methods		80

I3 Connection Of Project Management To Sigma   81

13. Managing The Sigma Risk Programme   82
I1 The Diagrammatic Representation Shown.   82
I2 O: Sigma Risk Management Decomposition   82
I3 O: Management of the Risk Programme   83
I4 1: Managing of the Sigma Process   83
I5 2: Facilitation of the Sigma Programme   84
I6 3: Practising the Sigma Methodology   84
I7 Relationship Of Dataflow Diagrams and SIGMA   85

14 Project Management Improvement   90
I1 Weakness In Project Management   90
14.2 Process Steps Required   90
I1 Collection Of Information   90
I2 Agreement On Resources   91
I3 The Score Graph   92

15.Remembering The Main Risk Management Points   92

Index   93 .

1. Risk Management: The Concept

I1 Programme Management

Programme Management may have many responsibilities, but the most important of all is the ability to identify and positively execute plans to manage the risks threatening the objectives.

Through a process of structured interviews and plans the Assessment Analysis is used to highlight the specific Events which may turn into Risks. During the interviews Assessment Analysis is used to capture the key Events from the interviewees.

In turn, the Assessment Analysis provides a life-cycle process, which highlights the primary prioritisation of the risks. In large, complex and critical programmes, it is essential that a true prioritised report is available so that the imminent threats can be managed first.

The process commences by identifying the most important events which may become threats to a project. These are given priority, support and management expertise. Once the prioritisation exercise is completed, the participating people are notified and subsequently interviewed to bring out and capture any possible concerns they may have.

Within a programme, projects are prioritised to ensure that those most critical to the programme's success are given priority to scarce resources.

I1 The Sigma Methodology

The Sigma methodology allows the capture of collective knowledge and expertise from those involved on the project, in a form that facilitates the communication of Events, Assessments and the pro-active management of Risks. Sigma can be applied to any type of project, or programme.

In essence, this is the mechanism by which the functions of programmes and projects are held together as a result of the principles operating within the *Sigma* methodology:

Σystematic: The varied Events, their Assessments and the consequential Risks relating to or consisting of a system. Methodical in procedures and plans, these are addressed to those involved and deliberating within the parameters of their systems development responsibilities. The results being dependable on:

Interaction: The mutual or reciprocal action which encourages those involved in the programmes and projects to communicate with each other and to work closely with a view to solving the threatening Events before they impact on the development of the system. The individuals involved maintain a -

Generic approach, which relates and characterises the whole group of those involved in assessing the Events and attacking the threatening ones before they become Risks to the development of the system. The end result being the avoidance of apparent problems within the pre-defined users systems requirements. This is enabled by following the -

Methodology: The system architects and the risk management practitioners simply follow the approved body of systems development methods, rules and management procedures employed by their organisation. For practical or even ethical reasons, it must be noted that with such a philosophy, it is seldom possible to fulfil all requirements of very large organisational systems. As such, the *Sigma* methodology is administered in -

Applications: Putting to use such techniques and in applying the Risk Management principles in the development of various *applications* will involve numerous and varied activities. A concrete issue in developing new applications is the problem of communication among the people involved, the motivation constantly needed for *generic* work, the ability to *interact systematically* and in using a structured systems *methodology*.

1.3 The Risk Management Cycle

The concept being a simple one as shown in the diagram below:

Risk Management Cycle

The **Sigma** (Σystematic Interaction and Generic Methodology for Applications) was developed by by the author whilst employed by **PsySys Limited,** over a period of some twenty years. The methodology was used for PsySys' international clients, from 1982 onwards. The idea of a structured approached to organisational problems proved beneficial to customers and users who integrated the full process with other methodologies, such as Structured Systems Analysis and Designing methods and Project Management procedures.

1.4 The Integration Of Methodologies

The comprehension of how to integrate the three methodologies can be achieved, simply by following the concept as shown on the following diagrammatic representation:

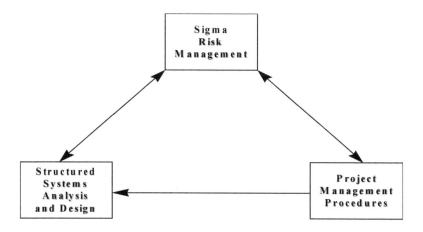

Integration Of Methodologies

1.5 The Inter-relationship

The various steps included in each of the methodologies are named in the next diagram. Or, to a further extent, the various stages of system development and the steps taken to manage projects and adopt the risk management cycle, are shown on the next page:

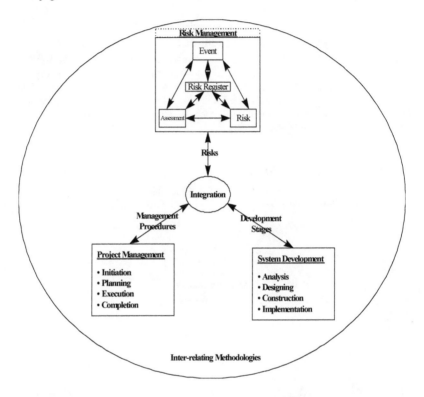

1.6 Managing the Risk Programme

It is basic business sense to identify, assess, manage and monitor risks that are significant to the fulfilment of an organisation's business objectives. In recent years businesses have been transformed by, and are in many cases heavily dependent on I.T.

The financial consequences of a breakdown in controls or a security breach are not only the loss incurred, but also the costs of recovering and preventing further failures. The impact is not only financial: it can affect adversely reputation and brand value as well as the business' performance and future potential.

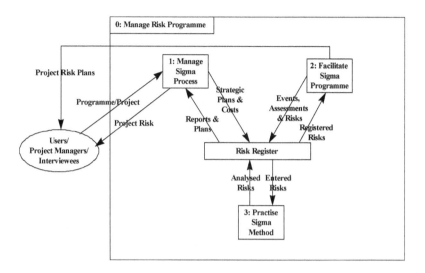

1.7 Impact On Business

Boards can regard inadequate system development as a significant risk, and where directors feel that this may be the situation in their organisations, they may need to ask tough questions of themselves and their management teams. Systems development and their risks is an issue that boards may need to recognise should regularly be on their agenda, and not delegated to I.T. technicians.

Business in the past was primarily confined to assessment of the risk surrounding fire, flood and Acts of God. In business today we have become high dependent on information systems. Failure to build computer systems as required by the users has a major impact on our business to function. The inability of companies to provide adequate systems can cause potential problems to customers, suppliers, employees and an all round havoc to information.

2. The Obligations Of A Corporation

2.1 Inter-dependency On I.T.

During the last forty or so years, the governments of the industrialised countries and the global businesses have become increasingly inter-dependent on Information Technology systems, to the point where many organisations would cease to function if these systems were unavailable.

Creating a strategy to ensure business continuity is, therefore, a prime consideration and whilst we are all familiar with the threat to the operations of corporations, such as system failures, badly developed and running systems, we are less aware of the potential risks to our I.T. systems and their costly development.

In the early days of I.T. few people envisaged the level of complexity and connectivity that would be achieved by the end of last century, particularly through the explosive growth in the use of the Internet. In the last ten years the number of the individuals and organisations that are connected to each other electronically has risen exponentially, thus increasing the risks of substantial loss through internal and external threats, some of which can paralyse distributed systems.

More and more transactions and communications will be carried out electronically and therefore the conditions to create risk free systems are essential to build confidence in information, communications and technology.

2.2 Internal Control

A further complication is that organisations are now subjected to a number of new laws and requirements that add further burdens to the hard-pressed executive. Specifically, these relate to the prevention of unauthorised access to information stored about the general public and employees, as well as allowing them to access information stored about them.

Furthermore, the importance that is placed on the protection of the information base of an organisation is emphasised by the rules on corporate governance being strengthened to ensure that protecting the electronic assets is taken seriously by Senior Executives.

Compliance with all of the influences that can affect the construction of new systems in your business or organisation is best achieved by the development of a methodical risk management practice. This includes the collection of events, the cataloguing of the assessments leading to the identification of threatening risks and the plan to assist in the execution of various tasks. Thus, securing the building of the I.T. system as per target, with the least possible risks.

To achieve such standards, the following elements should be included:

1. The concept of risks planning and their control to be managed by a Risks Management Practitioner,

2. A vulnerable part of the management of risks is the Risks Management system itself and the data stored in it,

3. The risks plans as reported by the Risks Management system is, also, of a significant importance, in evaluating the events and in the execution of actions.

2.3 Staff Training

One of the greatest threats in Risk Management is the non-acceptance of it or misuse of the system by your own personnel. Obtaining their buy-in is essential. It is obvious that someone is going to be appointed to ensure that the risk management principles are adhered to and the system maintained throughout the organisation.

It is, also, important that if information leaves your organisation by its transference to others for use on your behalf, that they have adequate systems to protect to your standards. The successful implementation of your Risk Management is reliant upon people and in particular your employees whose contracts of employment may need adjustment to protect the company and the adequate management and execution of plans for the solution to the threats caused by the risks identified.

To build a complete picture of what the Risk Management cycle includes, please refer to the diagram drawn below:

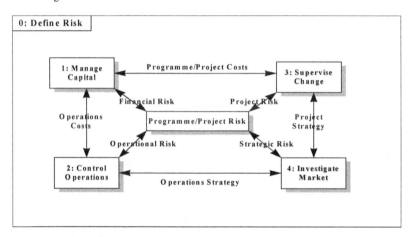

Programme/Project Risk: One Component Of The Total Business Risk.

Computer and Network Management ensure that we secure the processing facilities and generally prevent loss, modification or misuse of the Risk Management information entered on the risks systems. Computer based risks classification and control of the written risks policy allows management to have direction to control and monitor the information output by the risks system.

There are two ways in which you can approach the creation and implementation of risk management. By:

- Completing the process internally, or
- Utilising the services of external Risks Management Consultants.

3. Risk Management Analysis

3.1 Assessment Of Risks

Fundamental to the creation of a Risk Management system is the assessment of the risks (Risks Analysis) to your business and the potential loss that could accrue if things go wring. Risk Assessment software tools are available in the market, which can be used by consultants, or by internal staff. What is important is the ability to assess the risk to your business and the cost to protect it against the risk. The end result is that you have to make the valued judgement on the amount the business spends on the implementation and the monitoring of a risk policy.

Products and systems are available to counter the threats and risks that have been identified. There is a wide range of options available, but remember that anything chosen will require expertise to design and complete a system, taking into account how the various solutions will inter-react with each other. Like all things to do with I.T., the design and implementation of systems' risk solutions are only as good as the people installing them.

3.2 Communication

The most important factor in the success of any management style is the ability to communicate with each other, one to one or in groups of people. The art of communication is just as important to the whole process of the management of risks. More so where the risks identified have become a threat because of the problem of human communications.

This is where the appointment of an experienced and trained Risk Practitioner is worth the effort put into securing such individual/s.

3.3 Risks Management Practice

A trained Practitioner will have enough knowledge to run and maintain the system, as well as ample experience to be able to communicate with all levels of employees, hold meetings and ensure the plans executed.

In brief and as the diagram on the next page shows, the Practitioner will be responsible for the complete Risk Management cycle.

3.4 The Full Risk management Cycle

I.T. Risk Management

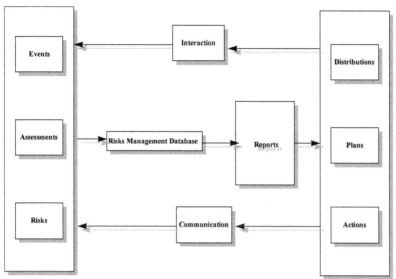

The Full Risk Management Cycle

In analysing risks, certain counter measures may have to be looked into. The mechanisms for safeguarding the construction of your information system is by managing risks and avoiding the threat of failing to build the required system.

4. Safeguarding Information Systems

4.1 Correct Software Control And Maintenance

Writing software is not a precise science and software companies are continually updating their software as information is received from users on their experiences.

They also become aware of the ways in which hackers are breaking into their systems and produce 'fixes' to prevent this happening. Regular monitoring of your suppliers' web sites and contacts with them will allow your organisation to keep up to date with the changes.

The mechanism for safeguarding your information systems include:

- The simplest and most important information security safeguard,

- Removal of known bugs,

- Updates with latest revisions,

- Secure 'back doors' into your systems.

I1 Access Control And Maintenance

Access control is one of the fundamental parts of the security policy as it defines the employees, customers, suppliers or partners who are allowed to access information stored on your computer system.

Achieving the balance between security and practicality takes time and deals with people's own perception of the right to access information and their perceived position in the organisation.

The following are a few steps which will help:

- Deter hackers through use of access screens and passwords,

- Authenticate users by use of passwords and/or token,

- Protect sensitive systems by special authorisation,

- Ensure passwords are secure,

- Revoke all passwords.

4.3 Firewalls

A Firewall is a security mechanism that allows limited access to your stored information, allowing approved traffic access according to a well thought out plan. A Firewall in its normal context prevents the passage of fire for a given period, but as stated above a Firewall in computer security does not prevent access in its totality.

Your security policy will have decided the traffic that is allowed in from the outside world on to your servers where information and programmes are stored.

A Firewall:

- Should be used on any system connected to the Internet,

◍ Can block communications likely to cause damage,

◍ Must be correctly configured and constantly monitored.

4.4 Vulnerability Assessment

The analysis of a network to identify points of weakness, if done manually from scratch, is an extremely time consuming operation. Therefore, software has been developed to complete this task. Experience in understanding the results is important to ensure implementation of the right solutions to cure any weaknesses.

◍ Network scanning:

  ◍ Looks at entire network from hacker's perspective,

  ◍ Looks for poor configuration and maintenance.

◍ Host scanning:

  ◍ Looks at individual systems to ensure technical standards and business needs are maintained

  ◍ Covers software configuration, passwords and auditing on individual systems.

4.5 Intrusion Detection Systems (IDS)

Intrusion detection is the art of detecting inappropriate, incorrect or irregular activity.

There is often a distinction between misuse and intrusion detection. The term intrusion is used to describe attacks from outside, where as misuse is used to describe an attack that originates from the internal network.

The purpose of an intrusion detection system is to detect unauthorised access or misuse of a computer system.

◍ IDS monitor intrusions into a network and detect potential attacks:

  ◍ Network IDS

    1. Detect attacks from outside,

    2. Must be carefully configured to avoid false alarms,

  ◍ Host IDS

    3. Can look for unwanted 'graffiti' on web sites

    4. Can detect attempts to steal passwords and data and tamper with audit trails.

I1 Cryptography

Cryptography has been used for centuries to provide a means of sending messages in code so that they mean nothing to the casual reader.

It was an obvious extension to use cryptography to secure data held on or transmitted by computers. As computers become more powerful it is necessary to increase the power of the encryption so that the codes used cannot be broken within a reasonable time scale.

Cryptography:

- Provides the most effective way of ensuring confidentiality, integrity and authentication of data,
- Should be used for secure transmission of data both internally and externally,
- Must utilise digital signatures for important electronic documents.

4.7 Virus Protection

'Virus' is used as a generic term covering any of the various forms of malicious software code that circulates freely on the Internet. Malicious code can cause immediate and obvious damage to a system, by tying up network resources or by destroying data.

Or, it can work in a more insidious way by causing cumulative deletions from, or insertion into, databases. Even more dangerously, it can open back doors into your system.

These may enable hostile outsiders to take control of parts of the network:

- Virus:
    - Covers any malicious software –whether damaging or not,
    - Circulates freely on the Internet, but can be on any floppy disc or CD,
    - Can tie up resources, destroy data, or change it subtly,
    - Can allow hostile outsiders when you have been attacked.
- Virus protection can:
    - Alert you to possible attacks,
    - Provide you with protection,
    - Give you solutions when you have been attacked.

4.8 Content Security

Content security/filtering is the management of the content of digital communications.

For legal reasons companies need to control the content of the material created, stored and forwarded on their networks. Content filtering software will permit scanning of e-mail and Internet sites for criminal, defamatory or racist material.

Such software must be carefully configured and companies should have a clear policy on its use.

The software should:
- Enable you to control the content of e-mails and Internet downloads,
- Reduce legal liability for illegal or damaging material,
- Be carefully configured and monitored,
- Enable a policy to be clear to all employees and third parties.

5. Business Related Overview

5.1 Business Responsibilities

There are now a host of legal and regulatory definitions detailing Board/Business Owner's responsibilities to ensure that information stored on computer systems, particularly with reference to people's personal information, be they employees or customers, is protected from improper access or use.

Corporate Governance defines Business Continuity and the management of the organisation's reputation as being the responsibility of the Board within public quoted companies. Compliance with the policies laid down is equally applicable to non-quoted companies, as it ensures best practice and safeguards all of our businesses. To comply, it must be shown that the risks to the organisation have been assessed and policies have been enacted to ensure that, as far as practicable, the potential for damage or loss has been reduced to realistic levels.

In the electronic age every aspect of our lives is becoming interdependent on others for computer-based transactions, especially in the business world. Around half of the interruptions to Business Continuity are now as a result of a failure in areas related to computers.

5.2 Corporate Governance

It is basic business sense to identify, assess, manage and monitor risks that are significant to the fulfilment of an organisation's business objectives. In recent years businesses have been transformed by, and are in many cases heavily dependent on I.T. The financial consequences of a breakdown in controls or a security breach are not only the loss incurred, but also the costs of recovering and preventing further failures. The impact is not only financial: it can affect adversely reputation and brand value as well as the business' performance and future potential.

Boards can regard inadequate system development as a significant risk, and where directors feel that this may be the situation in their organisations, they may need to ask tough questions of themselves and their management teams. Systems development and their risks is an issue that boards may need to recognise, should regularly be on their agenda and not delegated to I.T. technicians.

5.3 Business Continuity

Business in the past was primarily confined to assessment of the risk surrounding fire, flood and Acts of God. In business today we have become dependent on information systems. Failure to build computer systems as required by the users has a major impact on our business and how the company functions. The inability of organisations to provide adequate systems can cause potential problems to customers, suppliers, employees and an all round havoc to information.

5.4 Reputation Management

The management of a company's reputation is a vital issue for any business. It is the responsibility of directors to ensure that the issues are fully understood at board level and that time is spent putting in thorough, but flexible, plans to deal with them.

A good reputation can:

1. Put you above competition,

2. Attract better staff,

3. Attract business partners,

4. Attract investors,

5. Open doors to different markets,

6. Protect your business during a crisis.

5.5 Good Corporate Reputation

According to research carried out by MORI and Fortune magazine, the seven key aspects for a good corporate reputation are:

1. Financial performance, profitability and long term investment value,

2. The chief executive and senior management team,

3. Quality of your products and services,

4. Treatment of staff,

5. Social responsibility,

6. Customer service,

7. Ability to communicate.

5.6 Preparing For A Crisis

A crisis is an event that can damage the company, its product, service or reputation. Crises include natural disasters, man-made disasters, product recalls, sudden and serious financial problems, and closures and strikes.

Preparing for a future can appear a mundane task, but in essence is necessary for the safe-keeping of a company's reputation.

Companies should:

1. Identify and prepare a crisis management team,

2. Form a crisis management plan,

3. Remind board members that more is at stake than their own reputation.

5.7 Information Assurance

Information Assurance is becoming the term to denote the protection of information and information systems by ensuring their confidentiality, integrity and availability, as well as providing for authentication and non-repudiation.

A comprehensive strategy will assure those with whom you wish to do business electronically, that you have taken all reasonable precautions to ensure that your business is safe to connect to. In order to be able to prove that a contract exists between organisations both sides must be able to prove that the electronic communications have not been tampered with and you have to know that the person/organisation with whom you are doing business is who they say they are.

Information Assurance is the practice of developing and maintaining a secure, effective information risk management system that protects the information infrastructure as well as the data held within it. This is not merely good business practice, but for public quoted companies, a legal mandate, against which businesses must be audited.

5.8 Information Availability

Imagine a company being denied access of information (including its own) by the activity of others. This would be irritating at best, but for those that rely on time-sensitive information, it could be catastrophic. Let us face facts: today's purchasers have less and less time and patience. If they cannot reach their preferred supplier instantly, they will seek an alternative, if at all possible. So, a company's ability to provide specific, accurate information, on demand, will directly affect its success.

5.9 Interdependency And The Critical Information Infrastructure

The drive for greater efficiency in the global economy has created a situation where there is a highly sophisticated network of dependency on others and their systems, which enables us all to live and work. A failure on the system of delivery has a significant effect on the total operation.

The infrastructure interdependency, the onward march of information technology and the drive for more efficient business systems, demands an increasingly complex and sophisticated infrastructure, for the modern industrialised nation state. Systems such as transportation, power, banking, food distribution and telecommunications, to name but a few, are each crucial to a nation's well being in a highly competitive age.

Each of these systems, however, relies on some of the others for continued and efficient operation. For example, food distribution relies on transportation and telecommunications need power, conversely power needs telecommunications for the efficient control of its engineering operations. The reliance of each vital system upon one or more of its peers is known as infrastructure interdependency.

The more business driven and sophisticated a nation, the greater its vulnerability to infrastructure damage due to the compounding effects in a crisis of infrastructure interdependency. Whilst one cannot turn against the business drive for greater efficiency that enhances such vulnerabilities, it behoves governments to take an overview on infrastructure and insist on sufficient robustness to safeguard the nation, particularly in view of the potential exploitation of such weakness by terrorism and information warfare.

5.10 Internet Commerce

Internet Commerce is a key component of the Global Networked Business model, providing a company's customers and partners with end-to-end solutions to conduct business transactions and exchange of information.

As a background, probably the main driver for building the Internet Global Commerce model has been the need for continuous business improvement, leading to cost savings achieved by changes in the supply chain.

Global businesses and manufacturing techniques now ensure that the business works twenty-four hours, by seven days, by fifty-two weeks basis, which can only be effectively managed by the integration of computer systems on the equivalent global basis. The computer industry is reacting by introducing further interoperability between computer systems with the introduction of new software and bigger capacity technological devices.

Further on, the Internet, this network of networks stretches around the world and makes it possible for every computer to connect to every other computer. It is the biggest network in the world.

5.11 World Wide Web (WWW)

Often confused with the Internet, this is the most widely used means of publishing and accessing information that is stored on the computers, that are connected to the Internet.

The web allows you to link to many sites on the Internet. The basic concept is the page or collection of screens that it displays on your monitor. Within each page are links to related pages and other web sites. Each of these links is known as a hypertext link.

The web was originally used for text links only, but as it was further developed multimedia links were added. The web now contains pictures, audio and video links. With the addition of sound and graphics the web soon became the most popular way of linking to resources on the Internet.

Most organisations are worried by threats presented in using the web. Viruses are one of the main concerns, although these are unlikely to occur in normal surfing, except perhaps where Word and Excel documents are downloaded. Unauthorised access by outsiders is another.

Many organisations are also concerned about the employees wasting time surfing inappropriate web sites. To overcome this problem some companies choose to install a Content Security product that checks for ratings, profanity and legal disclaimers.

5.12 Business Case

The business case for justifying the expenditure on implementing a Risk Management system and the practising procedures to go with it in the early days of system development was more a leap of faith than a carefully evaluated financial case, and in many cases this may still be so.

Consultancy assignments and studies show that only a small percentage of organisations world-wide are taking this subject seriously. It is hoped that the information in this book will encourage you to review the Risk Management policy within your company.

5.13 Global Networked Business

A Global Networked Business is a company, of any size, whose networked infrastructure and use of technology speeds up the process of communication, and the sharing of knowledge-between prospects, customers, employees, partners and suppliers.

A Global Networked Business uses networks and information technology to:

1. Empower people to use and share information and to act more decisively,

2. Transcend traditional barriers – including geographic, financial or organisational barriers,

3. Increase responsiveness to customer needs and business opportunities,

4. Compete more effectively in the global marketplace,

5. Enables its customers, partners, employees and suppliers to access information, resources and services in ways that work best for them.

5.14 Implementation of a Risk Management Strategy

The success, or failure, surrounding a Risk Management Strategy depends almost entirely on people, those who are designing and developing it and those who are expected to implement it. If the system is designed in such a way as to be too complicated to understand and comply with, or in such a way that makes it almost impossible to do ones job, then it will be rejected by those who should implement it.

Implementation of the policy is likely to involve modification of employment contracts. Monitoring the passage of information in and out of the organisation will involve human communication as well as technological means of communicating, such as the analysis of e-mails sent by employees, which can infringe their human rights, as can monitoring which sites are accessed on the Internet.

Care in selecting those who implement the Risk Management principles can substantially influence the level of confidence attained. The programmes require parameters to be set and therefore the level of understanding of your business requirements and the software will influence policies success or otherwise.

The communication exercise to the employees is probably the most important part of the implementation. If left to the I.T. department, it may be delivered in seem-technical language or in terms of the needs of the business, rather than in terms and language to which employees can relate. Failure to allocate sufficient budgets to this area can put the success of a risk management policy in jeopardy. It is also important to remember to include training in this area as part of the employees' induction.

Adoption of a Code of Ethics can be a useful adjunct to the process, as can the use of an external communication company.

5.15 Monitoring

For reasons of corporate governance you are required to monitor the success or otherwise of your Risk Management process and to review at Senior Management/Board Level any significant shortcomings on a regular basis. It is recommended that in order to protect themselves and to ensure that the development of their systems are demonstrably risks free Senior Management should consider employing external monitoring and health check services.

Having implemented your Risk Management procedures and method of working, it is necessary to have a system review and reporting to ensure satisfactory communication among those involved and satisfactory compliance to the practice of risk management. The methodology will have been agreed during the specification and implementation process.

The review should include a list of all breaches of the system, the potential exposure because of the breach, actions taken as a result of the breach including the implications of the Risk Management.

It is the intention of this book to provide an insight into the subject of risk management together with sufficient background information to be able to enter into discussions on the subject, in order to determine the state of your policies on risk analysis. In addition to the financial and other benefits, one has also to consider the effects on the reputation of your organisation in the event that risks 'explode' into an uncontrollable situation.

6. The Sigma Methodology Explained

6.1 The Programme Objectives

It is a fact that most large, complex projects and programs fail to meet their planned objectives and as a consequence, most organisations are undertaking one or more aggressive programs at any point in time. These may fundamentally change the way the company conducts its business and failure to meet objectives on time may lead to a catastrophic loss of business.

Some projects or programmes can be chaotic at times. Objectives are evolving and plans and priorities are constantly changing. There is a temptation to accept this chaos as a necessary 'nature of the beast'. However, it is essential to move the programme forward in a traditional project management way by making sure that objectives and plans move forward.

Once we have clear objectives and plans, programme managers must control two fundamental factors if they are to be successful:

1. The business plan must be clearly identified,
2. The implementation of the program must be made explicit.

This can be answered by isolating the fundamental cause of most, if not all major project problems. It can be argued that projects only fail due to two fundamental reasons:

1. The plans are proven to be incorrect,
2. The significance of these plans are misunderstood.

The capture, analysis and communication of such assessments are, therefore, critical to the success of any project. This forms the basis of the Sigma Risk Management methodology. The Sigma method has been applied by PsySys to help many diverse organisations to deliver large, complex projects and programmes on time, to budget and in meeting the expectations of demanding users.

6.2 The Sigma Methodology

I.T. Risk Management

The focus of the **Σystematic Interaction and Generic Methodology for Applications** (*Sigma*) is based on he capture and analysis of the critical events and their assessments within the project plans, processes and procedures.

The *Sigma* methodology is essentially a framework process that allows the capture of collective knowledge and viewpoints from those involved on the project, in a form that facilitates communication of events, assessments and ensures the pro-active management of risks. This is accomplished by dramatically improving communications, risks are avoided or managed to the optimum and project objectives are delivered on time.

In essence, this is the mechanism by which the functions of programmes and projects are held together as a result of the principles operating within the *Sigma* methodology:

Σystematic: The varied events, their assessments and the consequential risks relating to or consisting of a system. Methodical in procedures and plans, these are addressed to those involved and deliberating within the parameters of their systems development responsibilities. The results being dependable on:

Interaction: The mutual or reciprocal action which encourages those involved in the programmes and projects to communicate with each other and to work closely with a view to solving the threatening events before they impact on the development of the system. The individuals involved maintain a:

Generic approach, which relates and characterises the whole group of those involved in assessing the events and attacking the threatening ones before they become risks to the development of the system. The end result being the avoidance of apparent problems within the pre-defined users systems requirements. This is enabled by following the:

Methodology: The system architects and the risk management practitioners simply follow the approved body of systems development methods, rules and management procedures employed by their organisation. For practical or even ethical reasons, it must be noted that with such a philosophy, it is seldom possible to fulfil all requirements of very large organisational systems. As such, the *Sigma* methodology is administered in:

Applications: Putting to use such techniques and in applying the risk management principles in the development of various *applications* will involve numerous and varied activities. A concrete issue in developing new applications is the problem of communication among the people involved, the motivation constantly needed for *generic* work, the ability to *interact systematically* and in using a structured systems

I.T. Risk Management

methodology.

I1 The Risk Management Cycle

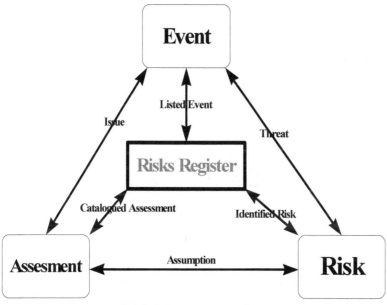

Risk Management Cycle

6.4 Features And Benefits Of The Sigma Approach

The key features and benefits of the PsySys approach are:

⦿ *Communication* — Provides a simple, common, language for the communication of risk up, down and sideways within the organisation, whilst avoiding the normal problems of political sensitivity and risk aversion.

⦿ *Control* — Enhances project control by exception management and achieves an overview of risk at senior management levels.

⦿ *Information* — Encourages the sharing of risk information, establishing common objectives, discouraging risk transfer and hence reducing the overall risk to all involved parties.

⦿ *Flexible* — An adaptable process which is rigorously applied to ensure that all significant risks are identified and controlled at the appropriate time.

⑩ *Acceptable* — The non-intrusive/non-bureaucratic management process improves management discipline across the organization and is readily accepted by project teams.

6.5 Assessment Analysis

The core of *Sigma* is the Assessment Analysis. This uses structured techniques to analyse project plans and identify the most sensitive events that are potentially unstable, and therefore the source of greatest risk.

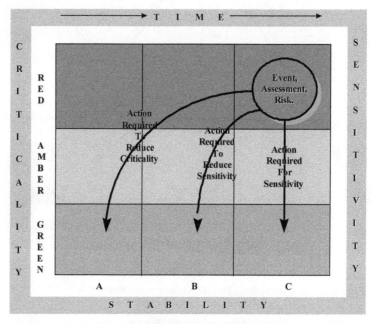

Project Criticality, Stability and Sensitivity: Measures To Be Taken To Reduce The Risk Impact

Everything is rated on a GAR principle: **G**reen, **A**mber and **R**ed scale; where G is always "good" and R is always "bad". This provides an instantly understood assessment on each stage: Events, Assessments and Risks in relationship with the time scales as used in the plans. This, effectively, provides guidance on how best to attack the risk.

6.6 Strategic Cost Analysis

Costing is a process within *Sigma* that can be used to define the cost of risk within a project or business area from as early as the proposal stage. It works by adding a 'quality' dimension to the estimating process so that high quality estimates, based on relevant experience, are treated differently from low quality estimates which are little more than guesses.

The output takes the form of a probability distribution diagram and a set of assessments which need to be managed in order to move the curve to the left and squeeze it (i.e. reduce the likely cost and the uncertainty).

Costing is particularly useful in the early stages of a project when the final cost of the project is subject to great uncertainty. The process has also been effectively used to define business budgets for re-structured business areas.

6.7 Risk Administration System Tool

A Microsoft Access based tool or any type of an ordinary spreadsheet can be utilised to allow the events, assessments and risks to be captured and reviewed by all stockholders in the program. In this way risks that would have been missed are captured through the identification of events.

6.8 Work Plan Analysis

Work Plan Analysis is a set of techniques that enables a rapid risk assessment to be undertaken on a complex project which is already in progress.

It is always difficult to focus on the right areas when the project organisation is large and the plans are extensive and likely to be multi-levelled. Using Work Plan Analysis, the 'poor quality' areas of a project are quickly highlighted for further investigation.

One very successful application of this approach has been through the use of Project Readiness Assessment Walkthroughs. These are structured review meetings held just prior to major project milestones or deliverables. Initially the project team explain their self-evaluation of the project status and are questioned by an independent review team. Potential risks arising are captured using the Assessment Analysis process.

6.9 Communicating The Risks

The Sigma techniques summarised above will only deliver its full benefits to any business if a suitable governance structure is quickly established to communicate the risk information and set suitable actions to mitigate the risks. The mapping of the process onto an organisation is the key step to ensuring that the investment in the *Sigma* process is fully realised.

6.10 How To Use The Contents Of This Book

The level of treatment of the material in this book assumes that the reader has already understood the principles already explained. With this in mind, the author attempts to explain the:

- Basic principles of risk management and contrasts the more traditional approaches with the theories that underpin the thinking behind the Sigma process,

- Practicalities of launching a Sigma process into a new project and/or organisation,
- Theory of Sigma and includes some of the practical considerations of applying the popular approaches of Project Prioritisation and Assessment Analysis,

⑩ Process of transferring ownership and embedding the method into the client's organisation.

7. The Principles Of Risk Management

7.1 Team Approach

An enterprise must escape from a culture based on transfer of risk between parties, to a team approach that is focused on attacking the real source of the risks. Methods must be effective without the need for detailed, time consuming analysis.

⑩7.2 The Definition of a Risk

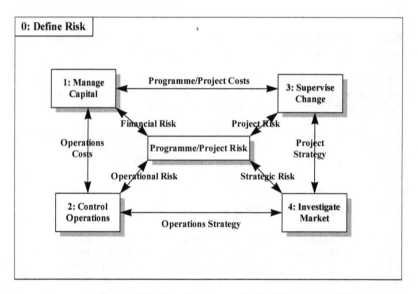

Programme/Project Risk: One Component Of The Total Business Risk.

A risk may be perceived as a possible loss. Risk is individual to a person or organisation because - what is perceived by one individual as a major risk may be perceived by another as a minor risk.

A risk is linked very strongly with competitiveness. Each decision has the possibility of resulting in loss. Each decision to introduce a new product into the marketplace can result in varying degrees of loss or gain. To be entrepreneurial is to accept risk, that is, the possibility of loss. A good entrepreneur's strength, however, is to make decisions which maximise possible gain. Hence minimise possible loss, which constitutes effective risk management.

Risk is inherent in all aspect of an organisation and may be viewed from four primary directions: financial, operational, programme/project and portfolio/products. Many risks are related to the running of the operations and its processes but is often in trying to change operations that the greatest risk is experienced. It is the management of risk in such 'change' projects that the Sigma methodology addresses.

A project can be described in its simplest terms as: Planning to achieve specific objectives and then executing the plans. The emphasis is on the word 'plan' as without a plan we have no project. So in the context of a project, a risk is something which might disrupt the plans such that the objectives of the project are not met. The discipline of Project Risk Management is thus a framework of techniques which allows the project manager to pro-actively identify and manage risks before they develop into problems which will impact the project plans.

7.3 Approaches to Risk Management

In recent years we have seen large projects in many areas of business suffering from a lack of control. The size of cost and time over-runs do not seem to be decreasing, despite the amount of management time which is being dedicated to analysing and quantifying the potential problems and selecting suitable personnel and processes. One may conclude that management, either do not have the correct methods and tools in place to attack the potential problems, or that they are not using, or do not understand, those which they do have.

In the early 1970's, the concepts of formal project risk management began to emerge. Hailed as the saviour of project managers, in practice the results have been mixed. Risk management has proved highly effective in certain mature industries - e.g. the Petrochemical or construction industry where project managers can base their estimates on years of similar engineering experience. Difficulties seem to be encountered when these traditional Risk Management methods are applied to innovative and fast evolving areas such as Information Technology.

7.4 Events And Risk Registers

Most projects will have an Events Register and some may have what they call a Risk Register. In effect, this tends to be a list into which anyone can input their concerns. It will contain references to current problems, questions, and assessments, difficult activities about which there is reasonable confidence and the odd real risk.

In any large project the Events or Risk Register quickly becomes swamped with items that require very different actions and many which do not require any action at all. All this leads to an inevitable loss of focus. Further, the content tends to be biased towards current problems rather than future potential problems.

7.5 Individual Interviews

One-on-one interviews can be an effective way of capturing risks. When people are not inhibited by management and peers, they tend to be far more open about their concerns. Unfortunately, most use very unsophisticated approaches such as "what do you see as your risks?" or "what keeps you awake at night?" Thus, if the person being interviewed is sensitive to discussing risks it may prevent the capture of any valuable information. At best the risks captured will tend to lack structure as they are not focused onto the future objectives that the project plans to achieve.

7.6 Group Brainstorming

Can be a very effective technique for opening up a very complex situation. However, information can be subconsciously suppressed by peer pressure, which may bias the discussion on one area at the expense of the rest of the project. Inevitably the mass of information captured is often difficult to focus, prioritise and allocate ownership.

In general, it should be remembered that the quality of the output is only as good as the quality of the input data.

7.7 Risk Analysis And Quantification of Risks

Risks may be difficult to capture reliably and concisely but further problems are likely to be experienced when trying to analyse them. Virtually all approaches to risk analysis are based on estimating the factored impact of the risk. This exposure to risk is a combination of the chance (probability) of an event happening and the consequences (impact) if it does occur i.e.:

Risk Exposure = Potential Impact x Probability of Occurrence

Fundamental problems arise when individuals are required to estimate, numerically, the impact and then predict (numerically) the probability. Estimates, which are often little more than guesses, result in a single point estimate of Risk Exposure, which is then given undeserved credibility in the detailed analysis of the risk and used as the basis for many major project decisions. Also, it is often the case that part of the risk impact can be quantified but often not the major part. An example can be based on an attempt yo quantify bad publicity, quality and relationship.

Some processes add complexity by rating the impact of risks in terms of financial, time scales, quality, performance etc., which quickly become very tedious to maintain.

7.8 Risk Control and Lack of Follow-through

Many risk management systems fail due to a lack of follow-through on actions. There is a surprising tendency to identify risks and then watch them happen!

This is caused by:

⑩ Failure to use the risk register to set appropriate action plans,

⑩ Lack of regular updates/maintenance of the risk register,

⑩ Absence of named owners and deadlines (lack of ownership),

⑩ Tracking generalities rather than specifics,

⑩ Concentrating on what can be done if the risk occurs rather than stopping the risk happening (pro-active),

⑩ Trying to transfer the risk elsewhere, without considering the consequences.

7.9 Risk Transfer

Risk transfer often occurs because the partner who knows most about the level of risk within the enterprise (i.e. the supplier/purchaser relationship) is encouraged to transfer this to the other partner. Once accomplished, the party with the most knowledge of the risk relaxes and the most ignorant partner inherits the risk. An example of this is the Purchaser insisting on a fixed-price contract in a poorly defined contract when they know that the supplier does not understand the scope of the contract.

The supplier then has a tendency to deliver the minimum possible and obtain sign-off for everything, irrespective of quality. The effect of this type of commercial 'table-tennis' is actually to increase the level of risk within the enterprise as the real risks pile up without intervention.

What is needed is a method that identifies and encourages the attack of real risk at source. Such a method would force projects within the enterprise to become pro-active by attacking risks, rather than waiting for events to unfold and then counting the cost, as recorded in the previous month's financial returns.

7.10 Risk Management vs. Project Management

There is often a tendency to treat risk management as no more than another necessary evil of project management. Thus, it often becomes an additional administrative burden for the Project Manager and consequently does not get the quality attention to make it work effectively.

In order to make risk management work, a shift in philosophy is required. This must lead the project team to view the process not just as another component of project management, but more as the communication stabiliser that holds the project together.

8. Basic Principles of the Sigma Methodology

8.1 The Process

The **Sigma** Risk Management method described in this book aims to provide an effective means of managing risks within all types of projects. The **Sigma** process grew out of a thorough assessment of the problems often encountered in project management and the techniques of the traditional risk management approaches that have been used to try and improve the situation.

Both good and bad principles were noted and new techniques were introduced to address key deficiencies. The resulting **Sigma** risk management process has a proven track record of delivering tangible results in large projects across a diverse range of organisations.

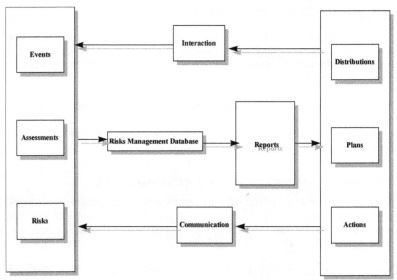

The Full Risk Management Cycle

8.2 Communication Of Assessments

As already highlighted, a fundamental reason for project failure is the lack of quality communications both within the project and between the programme and its environment. Most problems incurred by projects could be avoided if information was effectively communicated in a timely fashion. The problem is that there is so much information that it is difficult to decide what needs to be communicated and to who. This is where assessments come in.

Everything important associated with a project can be captured and tracked as an assessment:.

⑩ Activities are sized on the basis of assessments,
⑩ Milestones are set according to assessments,
⑩ Dependencies are based on assessments,
⑩ Plans are executed by making assessments.

Therefore, the capture, analysis and communication of assessments is critical to the success of any project, and this forms the core of the Sigma Project Risk Management process.

8.3 The Current Project Plan Is The Baseline

Risks are identified by capturing the critical assessments in the project plans are uncertain. In other words, whatever might stop the objectives, time scales and budget of the project plan being achieved. In this way, all assessments are effectively referenced to the project plans. Consequently, the plan provides the focus for the risk management process.

This approach keeps the risks specific, forward looking and ensures that the plan is always sufficiently detailed and up to date.

8.4 Uncertainty Equals Risk

Risk is inherent whenever there is uncertainty. The best judges of uncertainty are those who are asked to make estimates for the plans and, in most circumstances, the people who will actually have to do the work make the best estimators.

Combining this principle with the assessments captured from the project plans leads us to rate assessments for quality/uncertainty. Analysis is concentrated onto the areas of the project about which little is known and particularly the inter-dependencies that often represent the highest risk.

8.5 Judging the Quality using the *Sigma* Scale

To capture this vital information about how sure the estimator is, each estimator is asked, not only for assessments or the value of any estimate, but also, what quality he or she considers the assessment or estimate to be. This is not a judgement of the skill of the estimator. It is a self-assessment of the current quality of the basic information, upon which the project plans are based.

The *Sigma* scale is defined for multiple use throughout the process. It always means effectively the same thing i.e. A is always good and C is always bad. B expresses tendencies to the two extremes.

⑩ A (Green) means very good, high confidence, not important

⑩ B (Amber) means fairly good, reasonable confidence, not very important

⑩ C (Red) means very poor, little or no confidence, critically important.

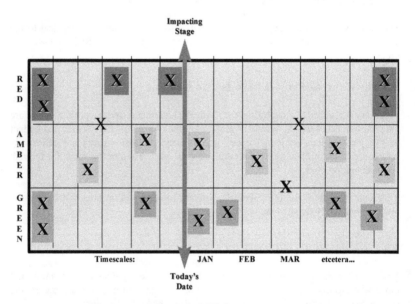

Impacting RIsks

It should be noted that the method does not allow the estimator to say that the estimate or assessment is of average quality. The whole principle is that we should be forced to make a choice between good, high confidence and bad, low confidence estimates and cannot 'sit on the fence'.

Using these simple A, B, C terms to express degrees of uncertainty, it is possible to encourage the estimator to reveal a wide range of uncertainty. Also, it is often possible to persuade him or her to make estimates when not normally prepared to do so. Being able to qualify an assessment or estimate with a C quality, often assures the estimators that they will not be forced into a given value or statement. We can thus gain vital information about the uncertainty and therefore the risks that may lie at the heart of the project without even asking for a risk.

8.6 *Sigma* Process Overview

The *Sigma* process consists of an integrated closed loop method, which logically progresses through:

- Project Prioritisation (for multiple project environments),

- Risk Assessment (consisting of Assessments Analysis plus Strategic Cost Analysis and/or Work Plan Analysis, if appropriate),

- Risk Prioritisation (to decide the 'order of attack'),

- Risk Control (to put the mitigation plans into action and monitor their effectiveness).

8.7 Project Prioritisation

A large organisation may have, at any one time, hundreds of projects of varying size, and nature. Yet many organisations have no formal mechanism for prioritising projects leading to problems such as:

- Not knowing which projects should be approved/resourced,

- Uncertainty as to which projects should be formally assessed for risk.Risk Assessment

Once the critical and potentially risky projects have been identified, Sigma offers three risk assessment techniques to identify and analyse the specific risks within each project.

The Assumption Analysis technique provides a backbone onto which the Strategic Cost Analysis and/or Work Plan Analysis approaches can be built.

In this respect Assumption Analysis would always be applied, Strategic Cost Analysis would be used in the early stages of a project or proposal to address the uncertainty in the cost/pricing of the project and Work Plan Analysis may be used to assess a very complex project which is well progressed.

8.8 Risk Prioritisation

The specific risks captured from each project risk assessment needs to be prioritised in order to allocate resources and decide the order in which the risks should be addressed.

Sigma provides a simple framework which rates each risk for Criticality, Controllability and Impact Timing. The resulting list of risks is captured in a Microsoft Access Risk Register (or any type of spreadsheet) and the risks are summarised in a diagram which provides an executive overview of the project risk

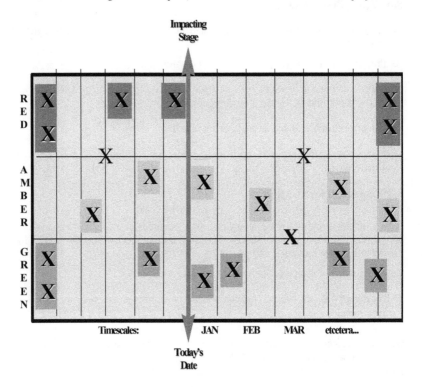

ImpactIng RIsks

profile.

The diagram shown in 8.5 is also appearing above, as this can be modified to include the list of impacting risks as reported by the Risk Register. Although this figure shows risks with the letter **X,** in reality, the **X**s can be replaced by the actual system generated risks reports.

8.9 Risk Control

Sigma provides a framework for risk control based on taking both strategic and tactical views of attacking risks. The strategic approach is achieved by applying trend analysis to the underlying assessments to identify any strong Risk Drivers, which can be neutralised together.

Tactical approaches match the complexity of the risk action plan to the complexity of the risk to minimise bureaucracy for simple-to-manage risks, whilst maintaining the necessary formality for complex risks.

The purpose of the diagram below is to show to the reader that at a certain point of time, measures have to be taken to reduce the impact of a Critical, Unstable and/or a Sensitive event, assessment and/or risk.

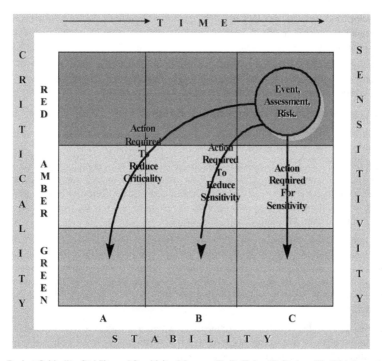

Project Criticality, Stability and Sensitivity: Measures To Be Taken To Reduce The Risk Impact

9. Initiating the Sigma Process

9.1 Organisational Culture

Some organisations are very risk aversive. This culture is often set unknowingly by senior management who suppress the open discussion of risks and create an organisation where stressing progress is the basis of management meetings. In this environment, nobody has any risks until they suddenly become problems!

In such situations, it may be necessary to re-sell the benefits of using risk management i.e. the worse thing that they can do is suppress risk, whilst the best thing that they can do is to identify risks and explain how they will be managed. Senior management need to be re-oriented from worrying about their current problems towards managing the risks within the plans that are designed to fix these problems.

If the organisation have any previous experience of using formal risk management techniques, it may be necessary to stress the particular benefits of a properly managed risk process. Effective risk management and plans:

⑩ Provide a common, apolitical language for the communication of risks,

⑩ Enhance project control by exception management,

⑩ Achieve an overview of risks at senior management levels.

The emphasis must be on fact that this method of non-intrusive, non-bureaucratic management process improves management discipline across the organisation and is readily accepted by project teams

In order to make the process work, the senior level sponsor must believe in the process and communicate with the organisation that they expect them to play a full role in the process. Buy-in must be both top-down and bottom-up.

9.2 Organisational Structure and Stability

When establishing the risk management process, the organisational structure and its stability needs to be considered.

Once the first round of risk identification interviews have been completed and the first Risk Review meeting is scheduled, it is essential to ensure that the meeting representation covers the risks being discussed and has the authority to take action. Often such actions require re-prioritisation of resources and this may mean that very senior representation is required to resolve conflicts.

The interview list and review meeting may be based on a project organisation which may change imminently. It is, therefore, always worth checking with senior level management on the likely organisational stability. If there is a re-organisation planned, do not remove people from the interview list on this basis alone. They may still provide valuable insights and certainly more than their replacement, if new to the project.

Perhaps one of the unstable environments in which to conduct a risk assessment is within a merger or acquisition scenario and it is worth noting that Sigma has proved extremely effective in this type of environment.

9.3 Effect Of Current Project Status

The stage and status of the project will have a strong bearing on the Sigma initiation process. Although all situations are unique, the following may give some helpful guidelines. Such a status may make a Strategic Cost Analysis very attractive. If there are no plans it may be difficult to undertake a full Assessment Analysis, but a high level assessment of issues or assumptions based on major milestones should be possible. Expect to identify a high proportion of risks that relate to missing or inadequate plans and resources problems. Projects just starting tend to have many events and few assessments (due to the lack of documented plans).

9.4 Project Fully Planned And Proceeding

Once the project has been substantially planned and is active, Assessment Analysis becomes the primary risk identification process. If the planning is very detailed and/or complex, some of the techniques of Work Plan Analysis may prove useful. For example, if there is a requirement to produce a first-cut of a risk register very quickly, for a rapidly approaching milestone, a Project Readiness Walkthrough would be an ideal approach.

9.5 Project In Trouble

The key aspect of a project in trouble is that it requires re-planning to put it back on track. Thus, the timing of the risk assessment relative to this planning process is very important.

If the re-planning process has not started there will be very little of the new approach to assess. It may be possible to influence this new approach by undertaking a risk assessment of the options being considered. To do this an Assumption Analysis of the alternative high-level plans can provide a useful framework for decision making.

If the project has been re-planned, then an Assessment Analysis of the new plans, possibly supplemented by Work Plan Analysis, is an appropriate way forward.

9.6 Interviews With Key People

Identifying the right people to interview is critical to producing a comprehensive and coherent picture of the risks facing a project. So, to decide on who should be interviewed, start with the project or programme organisational structure.

Depending on the scope of the risk assessment (i.e. single project, programme of multiple projects, portfolio of business projects etc.) it may be necessary to map the organisational hierarchy to ensure that the right people are interviewed and that the risks arising are reviewed at an appropriate level.

Working with the Programme or Project Manager, try to identify the 'key players'. A key player is someone within the programme/project who is likely to have either specific expertise in a particular area and/or insight into the environment in which the project is being implemented.

Key players tend to be Project Managers for a programme or Team Managers for a project with the addition of Users involved in the requirement capture and other activities. This group would likely form the initial interview list.

During the interview, these people should decide who else they need to participate. Interviewers need to exercise their judgement when evaluating the responses to this question. Typically it is necessary to go down at least one level below the Project/Team manager unless the team size is small.

One of the key features of the Sigma process is that of obtaining counter viewpoints within the organisation. Thus, the more people that are interviewed, the better. However, if many projects are being assessed for risk within an organisation, resource constraints will inevitably lead to reducing the interview pool. Under these circumstances at least two counter viewpoints must be obtained within each project. (e.g.: business manager and technical manager) so that the assessment ratings can be compared.

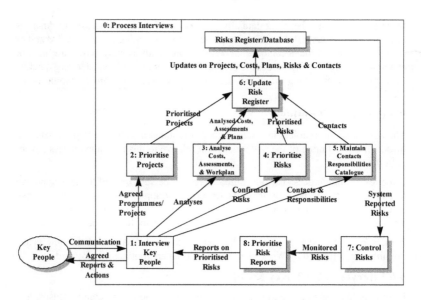

The Interviewing Process

9.7 Choosing A Suitable Risk Assessment Team

The team that will operate and manage the Sigma process requires a particular set of skills and background to be successful:

⦿ Experience of working in large (preferably non-consulting) projects and managing (preferably) medium sized projects (say 10-20 people).

⦿ Understanding of project planning principles and some exposure to associated tools.

⦿ Forceful personalities to ensure quality data captures in difficult client situations.

⦿ IT background, in order to understand the issues in IT projects and to help with using the Sigma support tools.

⦿ Some understanding of the clients business.

Note that it can sometimes be a disadvantage to have too much knowledge of the clients business in applying the Sigma process. This is because there may be a tendency for the interviewer to get into too much detail in non-risky areas and take too much of the client's time in the process.

9.8 Risk Review Meetings

After the initial round of interviews, a suitable forum must be established to discuss the risks identified. The client may suggest that the risks are discussed as part of the regular project meeting. This should be resisted unless the risks are an early agenda item and there will be sufficient time to get through the agenda with this additional discussion.

Typically, a minimum of an hour will be required for discussion of the risks - all if possible but just the most critical if not. If discussion of the risks is left to the end there will often be little time (or concentration) left to do the process justice. Also, it is likely that some events will end up being discussed twice. If discussed first, the risks tend to focus the meeting and get away from talking about progress onto the things that need to be discussed - i.e. what threatens the success of the project.

The best method is to establish a specific Risk Review Meeting with a representation consisting of the Risk Owners and chaired by the Programme Director, or the process 'champion' in the client organisation.

10. Project Prioritisation

10.1 The Project And/Or Programme

Only complex and critical projects need to have a fully structured risk management process in place.

A project is the controlled change of an enterprise. Projects are defined as having their own project plans and can be clearly distinguished from other projects (as opposed to being part of another identified project). For example, a rollout of a major system to individual countries with a specific plan for each country would only be identified as individual projects at a country level if they were effectively independent of each other.

The terms 'project' and 'programme' tend to be used very loosely. Sometimes the terms are used interchangeably or a programme is considered to be a large project. The definition preferred here is that a programme is a collection of inter-dependent projects.

From a risk management viewpoint the structure is important as it will dictate who is interviewed and where the key dependencies are likely to be. In addition, there may be a requirement to set up a multi-level risk management system where the project risks feed up into a programme risk register where there is a programme wide impact or the risk needs to be managed at the programme level.

10.2 Project Inclusion In Process

Basic criteria are required to decide which projects should be included in the project prioritisation process. Such criteria may be developed to suit the organisation but guidelines would include:

- Size and duration of project (i.e must be big enough to use significant resources for a significant period),
- Projects agreed by the business sponsor responsible for their implementation,
- Projects managed by the department/organisation and making use of their resources,
- Projects not managed by the department/organisation, but making use of their resources.

10.3 Prioritisation

The objective of prioritising projects is to ensure that the right projects have the appropriate resources and that appropriate management attention is paid to the most complex and critical projects. These projects are inherently risky, and in most cases will require a formal risk management process to support the other project management disciplines.

Therefore, there is a need for a process which:

- Produces a prioritised and agreed project inventory that can be used when allocating resources to projects,
- Helps management to focus attention on the critical projects which are inherently risky,

⦿ Highlights the projects which require formal risk management to guarantee timely delivery.

Project Size. An objective scale where A, B, and C are ranges of cost which are set by the business areas. The project is allocated to a cost range based on the anticipated total development and implementation costs over the life of the project.

Example, criteria could be:

⦿ C = More than ten equivalent resources,

⦿ B = Five to ten equivalent resources,

⦿ A = One to five equivalent resources.

Technical Complexity. How advanced or unknown are the engineering techniques which are needed to design and build the solution.

⦿ C = Highly complex, new technology, or scale never attempted before,

⦿ B= Some degree of complexity, some experience of proposed   technology or manageable scale,

⦿ A = Straightforward, familiar technology, small implementation.

Business Complexity. Many functions, areas of the business are involved in the project or impacted by the project.

⦿ C = Global system/involves many parts of the customer's business

⦿ B = Multi department/user groups

⦿ A = Single department/user group

Note that these scales are examples and must be re-defined to reflect the business objectives of the organisation under review.

10.4 The Project Prioritisation Process

10.4.1 The Standard Reference Matrix

In order to help refine the appropriate positioning of projects, each business area must identify two or three current key projects with which management are familiar. After agreement between the senior client and supplier managers, these will be positioned to make up a standard reference matrix. This matrix will then be used as the baseline.

10.4.2 Positioning and Prioritising New Projects

Initial positioning of the new project should be undertaken by the supplier and client project managers, together, as early as possible in the project. Complete consensus is not always possible at this stage, due to incomplete understanding of the business and technical issues involved.

However, it is essential that consensus is reached before any approval boards, or it is clear that the project is not ready to proceed. Lack of consensus can result in biased project ratings. Too high a rating will waste resources and cause frustration within the project team due to excessive management attention. Too low a rating will result in management neglect and consequently, increased risk.

10.4.3 Project Approval and Resources

In practice, the primary factor that is taken into consideration when approving a project is the Business Criticality. There are many projects that could be undertaken and only a finite resource pool available, so only the projects that will most improve the organisation's business should be approved.

Once management endorsement has been obtained, the results of the positioning exercise should be published to the teams and senior management.

10.4.4 When Projects Change Position on the CC Diagram

Once a project has been positioned it will be subject to review at least once every phase and certainly before progressing to the next stage. The most significant event in the life of a project is likely to be if it moves within either matrix.

A project may move because it is discovered that it is more or less complex than previously thought. It may have grown or been reduced in size. It may have become more or less critical to the client's business.

If a project moves, it is indicating something about the management approach required in the future. In particular, the approach to risk management should be reviewed and projects moving towards the top-right corner of the matrix should be assessed whilst projects moving towards the bottom left corner may be dropped from the risk assessment process.

10.4.5 Deciding which Projects Require Formal Risk Management

The position of a project on the diagram is also used to determine which projects should be assessed formally for risk going forward. The threshold can be set at any appropriate level. For example, some organisations work on the principle that any project judged to be a **C**, on either matrix, should be assessed for risk using the Sigma process. Projects falling below this threshold are still encouraged to use a formal method for risk assessment, but it is not mandated.

10.5 Risk Identification and Analysis

Sigma uses three windows; Assessment Analysis, Strategic Cost Analysis and Work Plan Analysis, to obtain a clear view of the source of any risk.

Although it is possible to create risk by poor planning and management, an element of risk is inherent in any enterprise. The main problem is to find a method of identifying the sources of the risks in time to allow specific actions to be taken to avoid or reduce the impact.

The nature of the project risks varies with the nature of the tasks, the size and the development phase of the project. Later phases may make use of methods for identifying risk based on the project's experiences to date. In early phases, good quality information and experience is scarce. Thus, different phases and different types of risk are best looked at by using different techniques and tools.

By using all these windows over the life of a project we increase our chances of identifying the major risks which lie within the project and to understand the nature of each risk. The process ultimately allows us to represent each risk in terms of:

- The potential criticality of the risk in terms of impact on the core project objectives,
- The point in time that the risk will impact the project if unresolved,
- The likelihood of the risk occurring.

Using these three analysis techniques; Strategic Cost Analysis, Assessment Analysis and Work Plan Analysis, it is possible to generate a single list that contains all the fundamental risks to the project. This list can then be used to decide which elements represent important risks and should thus be subject to proactive Risk Management by the creation and execution of dedicated Risk Plans.

10.5.1 Assessment Analysis

Fundamentally, projects only fail due to two reasons: either the wrong assessments were made, or the significance of the assessments was not understood.

The plans for any project are based on a few facts and many assessments. If the final out turn cost and time scales are excessive, it is obvious that either the assessments were wrong, or the estimators did not understand their significance.

It is important to note that these assessments are made due to the inevitable lack of total understanding or information which is inherent, particularly in the early stages of any project. Assessments are made to allow us to proceed with the project design or plans, but if they are important and prove incorrect, they will jeopardise the success of the project. Thus low quality assessments are at the heart of the risks in a project.

Assessment Analysis is the backbone of the Sigma methodology, and as such is at the core of Strategic Cost Analysis and Work Plan Analysis. It can be used at any stage in a project. However you should start the analysis as early as possible in the project to capture open events.

10.5.2 Identify the Sources of Assessments

Assessments come from a wide range of sources. At the beginning of a project all these sources are external to the project. As the project progresses it develops its own, internal assessments, while drawing on an increasingly larger set from its interfacing business and technical systems.

Assessments are captured by interviews with the key players in a project and its interfacing environment. Detailed interviews are conducted to kick-off the process and follow-up interviews are scheduled to track the status of the assessments captured and capture assessments over the life of the project.

Assessments may be explicit and be recorded in specifications, standards, plans or regulations. Many assessments will be implicit and only revealed by detailed questioning of the interviewee. Assessments may be made by the project, interfacing projects, the organisation or, they may relate to national standards and laws within the affected countries. Most assessments are made specifically to close open events, so that plans and estimates can be made and progressed.

At the start of a project there are likely to be many open events and few assessments. However, once the planning process is underway, the event are effectively closed by the assessments being made in the plans. Consequently, a project that is under way with reasonable quality plans will have few open events, if any. But, many assessments.

10.5.3 Prioritise Assessments

Many assessments made on a project are considered to be high quality as they are based on relevant knowledge or sound experience. Other assessments are insignificant when compared with the overall project objectives. Neither of these types of assessments are likely to be the source of risks and we therefore require a method to filter out high quality or unimportant assessments, so that we can concentrate on low quality or very important cases.

This risk identification method uses the knowledge of the key players in the interviews to assess two vital parameters which we have to know about assessments.

- The Sensitivity of the project to the assessments (i.e. How much does it matter if the assessments proves incorrect),
- The Stability of those assessments. How likely are they to change during the course of the project (i.e. How confident are we that the assessmentwill prove correct).

Sensitivity:

- A: would mean that the assessment being incorrect would have a minor impact on the project.
- B: would indicate a manageable impact.

Stability:

- A: would mean that there was high confidence that the assessment will turn out to be true,
- B: would be fairly confident,
- C: would mean a total lack of confidence in the assessment (i.e. it would almost certainly be wrong.

Using this method it is possible to conduct a systematic analysis of the assessments on which a project is based. We can represent this assessment on a Sensitivity/Stability diagram.

10.5.4 Convert Key Assessments Into Risks

Assessments are identified through employing Assessments Analysis, Strategic Cost Analysis and Work Plan Analysis. Essentially, the assessments captured which are important to the project (i.e. C Sensitivity) and are considered to be unstable (i.e. C Stability), represent significant risks to the project.

These risks require action to bring them under control. The most effective way of achieving this is by raising the risks to a suitable body in the organisation who have the power to make the necessary decisions, set the actions and make sure that they are followed through.

Communicating risks to senior management groups is always difficult as they are often not familiar with the detail of the project. Subsequently, the risks need to be presented in a concise, clear way which explains what is causing the risk (and therefore indicates what needs to be done) and what would be the consequences if the risk is allowed to impact the project.

Thus assessments are converted into risks using the form:

If (the assessment proves incorrect)

Then (describe the consequences to the project or business).

It is important to express the full consequences of the risk or it may fail the "so-what", test when submitted to senior management. A good principle is to describe at least the immediate impact and the ultimate impact on the project programme or business.

An alternative representation of a risk is to leave the assessments stated in the positive and add the impact. The advantage of this is that the risk can still be expressed in a positive sense. Some people find this very important.

If the Sensitivity/Stability ratings are AA, AB, BA or BB we have an assessment which is not crucial to the project and is likely to prove correct anyway. This does not require further action at this time but it should be re-assessed at regular intervals to make sure that the quality ratings are still good.

If another combination of Sensitivity/Stability ratings are assessed (e.g. BC, AC, CB etc.) these are called potential risks. Such assessments need at least to be monitored carefully for change.   Retain the data and re-assess these assessments with the originators at regular intervals to check that their status has not changed significantly.

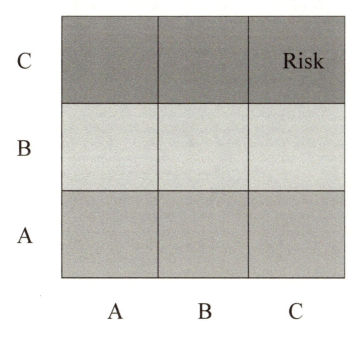

Table for Stability Sensitivity Ratings

It is important to remember that this assessment is only made at a point in time and will change as the project progresses. For this reason Assessments Analysis must be an ongoing process with regular interviews and review meetings throughout the life of the project.

10.5.5 Categorisation Of Assessments — Risk Drivers

When capturing assessments, it is often useful to identify what is driving the Sensitivity and Stability ratings allocated to it so that the source (or "driver" of the risk) is clear. This is simply achieved by categorising assessments into one of three types:

- **Policy** where the assessment relates to a business decision or policy, standards, resourcing priorities etc. The assessment requires management intervention to bring it under control.

- **Milestone** where the timescales of the activities are being 'squeezed' or timescale dependencies on other projects, suppliers etc. The assessment would be no problem if more time were available.

- **Technical** where the complexity of the undertaking is driving the ratings (e.g. untried design, hardware and software constraints, complex organisations etc.). The complexity is such that mistakes are likely irrespective of the time available.

Categorising assessments in this way can make the selection of appropriate Risk Plans easier by ensuring that the true source of the risk is addressed. Further, it can allow a strategic view of the risks to be obtained but only when a statistically large number of risks are being tracked.

10.6 Strategic Cost Analysis

Strategic Cost Analysis provides a means of assessing the cost risk in a project in its very early stages and commences the risk management process.

Strategic Cost Analysis is a process within Sigma which can be used to define the cost of a risk within a project or business area. It works by adding a 'quality' dimension to the estimating process so that high quality estimates, based on relevant experience, are treated differently from low quality estimates which are little more than guesses.

The output takes the form of a probability distribution diagram and a set of assessments which need to be managed in order to move the curve to the left and squeeze it (i.e. reduce the likely cost and the uncertainty).

Strategic Cost Analysis is particularly useful in the early stages of a project when the final cost of the project is subject to great uncertainty. The process has also been effectively used to define business budgets for re-structured business areas.

Its main aim is to highlight the possible range of final project cost, set contingencies, suggest a suitable Risk Budget and identify the fundamental uncertainties in the form of open events and critical assessments which can be converted into statements of risks.

The output of a Strategic Cost Analysis is a range of final project costs, the degree of risk in the project and open issues or assumptions that are linked to the poor quality estimates.

The main steps in performing Strategic Cost Analysis are to:

- Break up the project into modules and allocate ownership,

- Estimate time and/or cost of each module with the owner and attach quality ratings,

ⓞ Use a software tool to process the data,

ⓞ Analyse the results.

In essence the shape of the curve is dictated by the uncertainties (risks) in the cost estimates. Conversely, by managing these specific risks the curve can be compressed, increasing the certainty in the final project cost.

10.7 Work Plan Analysis

Work Plan Analysis may be used to focus on the risky areas of detailed, multi-level project plans when time is of the essence.

The final window for viewing potential risks to a project is through the detailed analysis of project work programmes and plans.

Work Plan Analysis is a set of techniques, which enables a rapid risk assessment to be undertaken on a complex project which is already in progress.

It is always difficult to focus on the right areas when the project organisation is large and the plans are extensive and likely to be multi-levelled. Using Work Plan Analysis, the poor quality areas of a project are quickly highlighted for further investigation.

One very successful application of Work Plan Analysis has been the use of Project Readiness Assessment Walkthroughs. These are structured review meetings held just prior to major project milestones or deliverables. Same kind of walkthroughs as used in structured systems analysis and design. Initially, the project team explain their self-evaluation of the project status and are questioned by an independent review team. Potential risks arising are captured using the assessments analysis process.

10.7.1 Quality Review of Plans

One would normally expect high quality detailed plans to be available for the current and next stage of the project, with the quality and detail decreasing progressively into the future. Thus the high level (Level 0 and Level 1) plans should normally be well defined with the definition becoming progressively lesser accuracy at the lower levels.

Remember that a high level of detail is not always equivalent to quality. It is possible to produce very detailed plans that are based on very poor quality assessments.

If the analysis of the plans shows areas of poor quality or incomplete planning in the current phase or at the higher levels, it is likely to indicate the presence of risks. These should be investigated for their root causes using Assessment Analysis.

An overview like this can be produced by using a dedicated checklist which is specific to the type of project being discussed. However, it can often be achieved by self-assessment from a number of the project's key players.

Further high-level views of potential project risks can be obtained from conducting a Contingency Analysis and/or a Time scale Risk Analysis using project management software packages.

10.7.2 Locate Hot Spots in the Plans

While project management packages are effective in producing likely cost profiles of the type produced by Strategic Cost Analysis, they are not very effective at identifying specific underlying risks within the project.

Skilled analysis of plans will reveal sensitive areas, which cannot be avoided. These Hot Spots may occur because of sensitive or multiple inputs to work packages, because of resource conflicts between packages, or through the difficulty of getting the output of the packages genuinely approved and agreed or multiple output which all rely on the completion of one work package.

The assessments in such areas of the plans may be analysed. Once identified as risks, the *If/Then* statement should be used to express the source and possible consequences.

10.7.3 Project Readiness Walkthroughs

Project Readiness Walkthroughs are a highly effective way of ensuring that all risks have been captured from the assessments analysis process. When used independently from, or prior to, an Assessments Analysis, the walkthrough provides a start to the risk assessment process which is particularly useful when time is short.

10.7.3.1 The basic objectives of a Project Readiness Walkthrough are:

⑩ Assess the readiness to meet milestones for high-priority projects,

⑩ Identify any corporate resource contention that may exist, and assist with prioritisation,

⑩ Identify common events and risks, and communicate information to support other project teams,

⑩ Help project teams achieve their goals by identifying and/or providing resources for assistance as needed,

⑩ Assess enterprise-wide risk through the evaluation of multiple projects.

10.7.3.3 The process follows a standard structure.

The project team are briefed and provided with a self evaluation form:

Ref.	Project Area	Pre-Milestone	Milestone	Post-Milestone
1.	Data conversion process			
2.	Data volumes/sizing			
3.	Operations changes			
4.	Systems changes			
5.	Test plans/test execution			
6.	Contingency plans			
7.	Staffing and resources			
8.	Controls and standards			
9.	Metrics and benchmarking			
10.	Training/education			
11.	Cross project communication			
12.	Customer impact			
13.	Problem reporting/escalation			

Form for Capturing Confidence Levels

10.7.3.3 The project team meets and fills in their confidence levels (A-C) against each category. The most pessimistic score is logged in each case to produce an entry matrix.

10.7.3.4 The Walkthrough is scheduled for (say) 3 hours.

10.7.3.5 The typical participants are:

⑩ The project team (key players),

⦿ Key players from associated projects,

⦿ An independent review team of senior management,

⦿ An independent facilitator,

⦿ Secretarial support.

10.7.3.6 The Walkthrough starts by concentrating on the Cs to find out what is driving them. Appropriate assessments and risks are captured in real time.

10.7.3.7 The Walkthrough is completed by going through the As and Bs and asking for them to be (briefly) justified. Any additional assessments or risks are logged.

10.7.3.8 The matrix is updated to reflect the risks identified to produce an exit matrix.

10.7.3.9 The risks reviewed regularly and actions stated to ensure that they are brought under control.

10.7.4 Pre-selected Assessments

This is a process very similar to the Walkthroughs, where the assessment criteria are a set of pre-selected assessments. A set of high-level assessments are defined on the basis that this is what is needed for the project to be successful. These assessments are then agreed with the sponsor before undertaking short focussed interviews with as many team members as possible. In essence, the team members are asked for their opinion on the Sensitivity/Stability of each assessment.

The results are collated to show trends across the team. The strength of this approach is based on the teams perspective and not on what the reviewer assumes.

10.8 Risk Prioritisation

Prioritisation allows the Project Manager to direct limited resources at the most critical project risks.

The objective of the risk prioritisation is to identify the most significant risks out of all those which have been identified by the various analysis methods used. Once all the risks have been collected in a consolidated list or register, they should be placed in order of priority and attacked via a logical, planned programme. The problem is to decide how to place risks in an appropriate order.

10.8.1 Assessments and Risks Register

All assessments captured should be held in an Assessments Register. Only critical assessments will be converted into risks and held in a Risk Register. This is done by filtering the assessments and consolidating them into risks. All information captured will be rationalised and details of their source and consequences will be traceable.

10.8.2 Positioning Risks

The primary criteria used for prioritising risks in Sigma are:

⦿ Criticality

⦿ Timing

⦿ Controllability

I.T. Risk Management

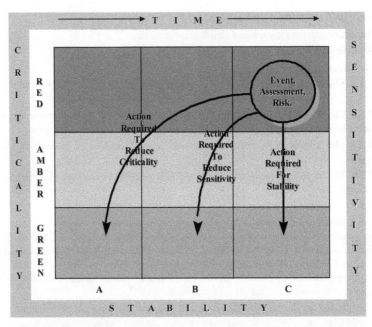

Project Criticality, Stability and Sensitivity: Measures To Be Taken To Reduce The Risk Impact

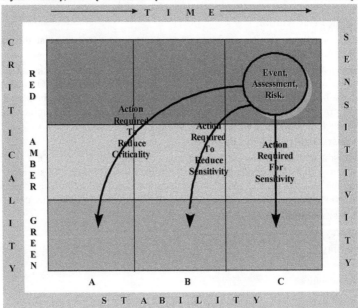

Project Criticality, Stability and Sensitivity: Measures To Be Taken To Reduce The Risk Impact

Criticality

In certain instances the risk may undermine the basic objectives of the project and no amount of money will save the project if such a risk impacts. If not resolved, the uncertainty may halt the progress of the project. Such a risk may be related to the overall programme, a part of the programme, an individual part of the design, or even a particular module of software. This, also, provides a way of representing the effect of such risks on the overall project, where cost impact is small or meaningless.

To satisfy this need a Criticality index is defined. Criticality is in effect a multi-dimensional risk impact rating. Once again, we use the assessment A, B, C. C being dangerous, while A impacts the edge of the system design or the programme. Something which may be important in itself or to one group of users or designers but will not stop progress on the rest of the project.

In most applications of Sigma, Criticality has been the primary means of prioritising risks and has been described in terms of traffic-light ratings of Red, Amber or Green. This can be very effective at concentrating on the project impacts and/or avoiding confusion with ratings of assessment Sensitivity.

Positioning in time

One of the most important things we need to know is *when* we have to do something about an event, assessment, or risk. In the case of a project we need to know when the risk will start to impact the work. Then we must determine when we have to take action to prevent it happening or to reduce its impact. The timing of a risk should always equate to the latest time to start the first necessary action. In this respect, it is analogous with trying to stop a cancer. This must be done at the point that it starts to grow, not when it can be first be seen.

Controllability

Controllability is a measure of confidence that the risk will be managed. It should not be confused with the probability of the risk occurring, which is a measure of how likely the risk is to occur if nothing is done. The controllability grade cannot normally be assigned until the risk has been reviewed and discussed by senior management, whereby an agreement is reached as to their confidence that the risk can ultimately be managed. A 'C' grade means that no risk plans are in place and no action has been taken, whereas an 'A' grade indicates that risk plans or actions are well under way with a very high confidence of success.

10.8.3 Risk Register Reports

Impact Diagrams provide an overview or risk profile of the project. However, the detail of the risks is required for the risk review meeting in order that the detail of the risks can be seen, discussed and actions taken.

The order of the risks in the report is important so that senior management can focus on the key risks first.

If the impact diagram is used to prioritise the risk register the time element can be easily included. For instance there may be an urgent AMBER criticality, C controllability risk that needs attention that is not an obvious priority if the Risk Report is prioritised by Criticality and Controllability alone.

In essence the easiest way to prioritise is to use the Impact Diagram and to treat the highest priority risk as the one nearest to the origin, the next nearest being number two and so on. The intention is to order the risks so that they are roughly in the right order.

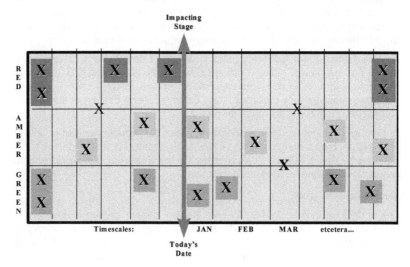

Impacting RIsks

10.9 Risk Control

Risks may be attacked at both the strategic and tactical levels. Strategic approaches look for trends and underlying causes for groups of risks. Tactical approaches take each risk at face value.

10.9.1 Strategic Approaches

A strategic viewpoint is achieved by using the Risk Driver approach. Each assessment and subsequent risk is categorised into Technical, Milestone, Decision or Resource to reflect what is driving the poor quality ratings:

⑩ **Technical** relates to assessments of a technical or complexity nature (e.g. the pure complexity of providing an interface)

⑩ **Milestone** which applies to assessments regarding timescale dependencies on other projects, or external suppliers, and can be used to identify linkages to project milestones (e.g. an activity which is not inherently complex but may not be feasible due to timescale constraints).

⑩ **Decision** which is used to describe assessments which require business decisions, business policies or standards (e.g. organisational announcements).

⑩ **Resources** which relate to resource deficiencies or priorities (e.g. insufficient training resources).

Note that with this categorisation, Policy has been split into Decision and Resource categories. This is necessary when the resource constraints are external to the programme/organisation. When the resource constraints are internal, a simple Policy Driver will cover this.

Categorising assessments and risks in this way identifies the main risk drivers, can simplify the identification of trends and assist in the development of appropriate risk plans.

For example, Red Decision and Resource risks are "show-stoppers" which generally require senior management action. Red Milestone risks are an indication of how tightly the programme is being "squeezed" and Red Technical risks indicate the complexity of the planned activities.

The Risk Driver chart indicates where particular effort is required. In the example below the relatively high number of Decision based risks suggests that the project is being put at risk by having to wait for decisions, probably from within the organisation. A steering committee meeting could potentially

Technical	Milestone	Resource	Decision	Total

resolve most of these risks. If the project is in the early stages, it is already showing signs that the time scales are too ambitious by the high number of milestone risks.

Red	7	15	4	19	*45*
Amber	12	19	15	26	*72*
Green	10	6	12	12	*40*
Total	**29**	**40**	**31**	**57**	**157**

It is important to note that noting that the "normal" Risk Driver profile trends with the phase of the project. i.e. "Normal" would be (using relative terms):

Phase	Technical	Milestone	Resource	Decision
Start of project	Rising	Min.	Max.	Max.
Middle of project	Max.	Rising	Falling	Falling
End of project	Falling	Max.	Min.	Min.

10.9.2 Tactical Approaches

Most risks will need to be addressed specifically (i.e. one action plan for each risk) to address the underlying assessment. Assessments that are placed in the C area of the Sensitivity/Stability matrix are unreliable, represent significant risks, and it is dangerous to continue with the project without taking action. We must do at least one of two things:

⑩ Stabilise them by escalating the assessment to senior management and obtaining agreements that increase confidence that the assessment will turn out to be true.

⑩ Make the project less sensitive to the assessment i.e. Reduce sensitivity by redesign, re-planning or having acceptable fall-back plans in place.

The actions taken to attack the problem may be very different, depending on whether we are trying to reduce Sensitivity or Stabilise. It is normal to try to Stabilise the assessment first before trying to handle sensitivity, which is usually more difficult.

10.9.3 Following Through on Actions

Just as it is possible to introduce risk by planning a project badly, it is possible to address many risks by using normal project management methods, as long as the risk has been identified early enough.

Risks generally come from two areas activities which have not been planned adequately or planned activities which are likely to go wrong. Therefore, the first filter to apply to the risks is to identify those which can be tackled by just improving the plans. Obviously, if there is no project plan the risk is unbounded and the first action must be to create a plan.

10.9.4 The Need for Risk Plans

Risks which cannot be resolved by improved planning must be tackled individually through the use of dedicated Risk Plans. Ultimately, the plans for reducing the risks must be incorporated into the main project plans.

Risk Plans may be divided into "Simple" or "Complex", irrespective of the potential impact of the risk. Simple means that it is possible to resolve the risk quickly say by a simple phone call or single task. For such risks, monitoring the status on the Risk Report is sufficient and minimises bureaucracy. Complex risks require significant resources and time to resolve them and for these a formal Risk Plan is required.

10.9.5 Essential Components of a Formal Risk Plan

For Complex risks, it is essential that a structured plan is formulated. This will clarify thinking, provide the necessary visibility and feed directly into the main planning process:

The basic components of a Risk Plan are:

⑩ The risk statement and its ratings,

⑩ Risk Owner and Risk Action Manager,

⑩ The driving statement and its ratings,

⑩ Assessment Originator,

⑩ Objectives of the Risk Plan i.e. Stabilise the assessment,

⑩ Criteria i.e. how will we know when the objectives have been met,

⑩ Risk Plan summary i.e. what are the steps to meet the objectives,

⑩ Reference to project plans,

⑩ Additional resources required,

⑩ Monitoring process i.e. how often and by whom,

⑩ Re-assessment of the driving assessment (completed after execution of the Risk Plan)

⑩ Fall-back plans i.e. what we do if the Risk Plan fails.

10.9.6 Devising Risk Plans to Attack Risks

10.9.6.1 Developing Risk Plans

I.T. Risk Management

It is preferable to develop a number of alternative Risk Plans before a decision is made to implement any one of them. The analysis technique used to identify risks can be used to help understand what type of plans may be appropriate. There are no firm rules that govern the way we approach Risk Plans but, if a simple logic is followed, it becomes easier to be sure all the possible strategies have been considered.

10.9.6.2 Plans for Attacking Assessment Based Risks

Two approaches should be considered for each critical assessment:

- Stabilise the assessment to ensure that it will be right,

- Reduce Sensitivity to the assumption, and make it matter less.

It can, also, be useful to analyse the Risk Driver to understand what sort of assessment is being considered:

- Technical assessments which may relate to the complexity of the design, implementation and the technical interfaces with other systems and projects.

- Milestone assessments which are based on timing constraints and co-ordination with other projects. Appropriate Risk Plans normally result in re-planning and the addition of new milestones to the project plans to check that external deliveries will be met.

- Policy based assessments (including Decision and Resource Drivers) which are caused by the business environment, its policies and the decisions or lack of decisions. Risk Plans aimed at stabilising the a assessment are normally appropriate (e.g. gaining high-level commitment to a policy or standard).

10.9.6.3 Plans for Attacking Planning-Type Risks

A Work Plan Analysis of the project plans may reveal risks that are related to:

- Areas where the plans are of poor quality (C), where we should expect good quality,

- Areas where high probabilities that the contingencies will be used have been identified,

- Areas where large contingencies have been added.

Risk Plans addressing risks that have been identified by an analysis of the project plans often involve the same type of process as that used in main project planning:

- Detailed (re)planning of sensitive areas,

- Re-timing modules and activities,

- Re-structuring the work breakdown,

- Re-allocation of responsibilities for elements of the work within the project,

- Exporting certain elements to other agencies,

- Forming sub-projects,

- Constructing models and prototypes,

- Resolving resources,

- Special programme control procedures,

- Planning parallel activities,

- Identification of linked milestones in other projects,

- Monitoring external dependencies.

10.9.6.4 Selecting Particular Risk Plans

The selection of a particular Risk Plan is normally a simple process; as the plan which is most likely to succeed is often obvious. Occasionally it is necessary to analyse the Return On Investment (ROI) of each plan in order to make a decision.

Having generated a number of alternative Risk Plans for each important Risk, we now have the task of deciding which plan to put into operation. It will normally be obvious which of these plans should be employed first. Our own experience and judgement will be enough, in most circumstances, to choose the plan on the basis of simplicity, cost, time to implement or chance of success.

Often, some selection criteria for Risk Plans will be used before the plans are prepared and this will obviously have a bearing on the plans proposed (e.g. on a tight time scale, only plans which can be implemented quickly will be considered).

Circumstances may change, however, and it is a good idea not to limit plans to those that may be appropriate at the time of initial selection. After all, the first plan that is tried may fail.

When we consider the merits of various Risk Plans, there are a number of factors that must be assessed:

- The time available before the last action date by which the risk must be reduced to an acceptable level,

- The duration of the proposed plan,

- The cost impact of the risk, which may limit the allowable cost of the Risk Plan to produce an acceptable rate of return,

- The remoteness of the risk, which also affects the minimum acceptable rate of return required of the plan,

- The collective judgement in the project team will have to be used to assess many of these factors. However, cost and time scales factors can be approached in a more structured manner, if the choice of plan is not immediately obvious,

- When considering how much to invest in a Risk Plan it is important to know how much time is available before the risk may impact the project. If the risk is a long way away, and remote from the heart of the project, we should think very carefully before immediately implementing a Risk Plan which will expend a lot of effort.

The expected Return On Investment which may be required before the project manager sanctions a Risk Plan, tends to vary with time. Experienced project managers expect returns as high as 50 to 1 if the risk is beyond the planning horizon.

The reason for this is that:

- Any Risk Plan contains its own risk that it may fail,

- Remote risks may disappear of their own accord,

- The fact that we are aware of a remote risk may allow us time to modify our plans to avoid them, without taking any special measures,

- The risk may not be as big as it looks at long range,

- It may be wise to find out more about a risk with a modest plan before we launch the full attack,

- If a modest plan succeeds we have saved resources,

- If a modest plan fails we still have time to launch another, more aggressive, attack,

- If an expensive plan is launched and fails we may not have the time or resources to carry out a further plan,

Error! Reference source not found. Also, as a risk moves through the planning horizon - i.e. when the risk is still there after detailed plans have been developed, it is essential that suitable Risk Plans are prepared quickly to maximise the Return on Investment. Time becomes the essential decision driver.

As the risk approaches, the acceptable rate of return drops rapidly, until plans with very low Returns on Investment may have to be implemented in order to save the project. If the risk is not high Criticality, it may become more cost effective to accept the impact in terms of cost or loss of functionality, and perhaps launch a sub-project later.

However, in these circumstances, we either failed to identify the risk in time to act, or we finalised the Risk Plan too late for it to be effective.

10.9.7 Project Risk Organisation And Responsibilities

The definition of Risk Management roles and responsibilities is crucial to the success of the Risk Management process.

Each risk needs a Risk Owner and every Risk Plan must have a dedicated Risk Action Manager, to ensure that the risk is properly attacked. Each of these roles should have published responsibilities. The roles may be full or part time, as appropriate to the size complexity and criticality of the project. The individuals may be drawn from within the project team, from either the supplier or client personnel, or from an outside source, if this is appropriate or the resources are not available within the project.

10.9.7.1 Risk Ownership and Risk Action Managers

It is common practice in traditional risk management methods to assign a single Risk Action Manager/Owner for each risk. This is normally the person who knows most about the area in which the risk would impact and also actions the risk.

The main problem with this approach is that, particularly on large projects, risks stop being identified effectively as the risk Owner/Manager of a risky area becomes overloaded with Risk Plans.

To avoid this situation, it is recommended that, on all but the smallest of projects (i.e. less than 10 people in the combined supplier/client project team), separate Risk Owners and Risk Action Managers should be assigned for each risk.

The allocation of separate tasks to Risk Action Managers and Owners, whenever possible, produces a system of checks and balances.

I.T. Risk Management

The Risk Owner:

- Sets the overall objectives for the plan,

- Certifies that the Risk Plan selected will work,

- Certifies that progress is on schedule and that the plan has succeeded in reducing the risk.

The Risk Action Manager:

- Plans the detail,

- Applies for any necessary resources,

- Carries out the plan and ensures that time and effort is not wasted on activities not strictly associated with the specified Risk Plan.

The Risk Action Manager will tend to be, naturally, efficient at such tasks as they will be busy with their own part of the project.

Separating the Owner of the risk from the Risk Management task is thus very desirable on large projects and ensures that Risk Plans do just enough to effectively attack the risks. It relieves the Project Manager of much of the monitoring effort associated with controlling a large number of different plans and permits him or her an overview of the risks to the project and the Risk Management process.

10.9.7.2 Selecting Risk Action Managers

The choice of Risk Action Manager depends upon the task in hand. Many risks can be managed by quite junior staff. Motivated systems engineers and administrative staff can be empowered to carry out a simple plan which may have very significant effects on the project (e.g. Collecting the latest information from Policy making bodies).

Using junior staff is very cost effective, it is very good training and it can be a very effective motivator. Allocating junior staff tends to prevent the Risk Plans dominating other routine activities, and thus disrupting the rest of the project.

Risk Action Managers should be drawn from outside the project for external risks. In this case the Project Manager identifies suitable candidates, but the project Steering Committee or advisory board appoints them. One of these groups is also responsible for ensuring that any external Risk Action Managers are correctly empowered and controlled by their own line management.

10.9.7.3 Selecting Risk Owners

The choice of Risk Owner is critical to the success of a plan. A person must be chosen who is:

- Knowledgeable about the project area that the risk will impact,

- Informed about the nature of the risk,

- Motivated to see the problem solved,

- Able to see when the plan has been successful,

- Able to recommend that the plan should be stopped.

Care is required to ensure that the Risk Owner is suitably senior and first-line reports to the Project Manager are normally appropriate. This is because risks will normally be reviewed by the senior members of the project team and it is essential that all key risks can be discussed in such a forum.

It is perfectly possible to have an external Risk Owner. The only problem with this external arrangement is ensuring that there is sufficient motivation in supporting the project which will be affected by the risk.

10.9.7.4 The Risk Review Board

The focal point of the Risk Control process is the Risk Review Board. This body ensures that the roles and responsibilities are set, the risks are understood at the appropriate level in the organisation, action plans are set and followed through and the communication flows at the senior level in the organisation.

The basic terms of reference of a Risk Review Board are:

Representation

The senior managers (e.g. Project Managers in a Programme) normally form the core of this meeting. In addition any (senior) Risk Owners will be required to attend. The meeting should be chaired by the Programme/Business Director and facilitated by a Sigma Risk Management expert.

Before the meeting

The Sigma Risk Team will undertake interviews of all agreed key players to capture important and risky assessments. All data will be captured in the Risk as database from which the Risk Register and report will be produced.

The Risk Register will be circulated to all Risk Owners one day prior to the meeting. It is the responsibility of each Risk Owner to review their risks before the meeting and to make sure that they fully understand the risk, agree with the ratings and are clear about any actions underway. To achieve this it may be necessary for the Risk Owner to communicate with the Assessment Originator and/or the Risk Action Manager.

In addition, each Risk Owner should review the Assumption Register (or relevant part of it) to check agreement with other people's assessments and ratings. Any major disagreements should be raised in the Risk Review Board. Minor queries should be fed back to the Risk Team.

During the meeting

The risks will be discussed in priority order as indicated by the Impact Diagram. The process for each risk will be:

⑩ The Risk Owner will clarify the risk briefly (if necessary),

⑩ The meeting will endorse or change the ratings for Criticality, Controllability and Impact Timing.

⑩ For new risks, suitable action/s will be agreed and a Risk Action Manager will be assigned to undertake the action plan.

⑩ For existing risks, the Risk Action Manager will report on mitigation progress and when the risk is considered as resolved, the Risk Owner will agree the closure proposal. Note that the risk is not formally closed until the Assessment Originator agrees to down-rate or close the associated assessment.

At the end of the meeting there will be time to raise any urgent assessment queries and discuss them as risks, if appropriate.

After the meeting

The Risk Register and Impact Diagram will be updated from the meeting and re-issued, within one week, to all Risk Owners as a formal record of the meeting and actions agreed. Risk Action Managers are then responsible to undertake the actions agreed in the Action Status section of the Risk Register, while Risk Owners are responsible for making sure that actions are followed through in a timely fashion.

Immediately following the meeting, the Risk Team will commence the next cycle of interviews.

10.9.7.5 Running Risk Plans

Once the Risk Plan has been agreed and approved, the running of the Risk Plan should be straightforward. The plan should be followed and progress reported to the Risk Owner and Project Manager against time and the resources expended.

Simple action plans (e.g.: phone calls, one day investigations etc.) should be noted quickly and tracked in the Action/Status field of the Risk Report. Complex Risk Plans which extend over a longer period (say more than 5 man-days) should be detailed in the Risk Plan section of the database and ultimately incorporated into the main project plan.

The decision on whether to incorporate a Risk Plan into the main project plans is a judgement based on the relative duration of the Risk Plan relative to the main plans. For example, there is little point in including a 3 man day Risk Plan in the project plans if the planning has only been carried out down to 20 man day activities.

10.9.7.6 Closing Risk Plans

There is a significant difference between closing a plan and just stopping it. Risk Plans should be stopped if they are not succeeding. Closure is only authorised when the objectives have been met. It is critical that Risk Plans are closed or stopped at the appropriate time in order to avoid wasting effort.

Further on, there is no point in expending effort on identifying risks, deciding which to attack, formulating alternative Risk Plans, and then conducting them, if the plan is not properly closed when it is complete or has failed.

10.9.7.7 Stopping a Risk Plan

A Risk Plan must be stopped when:

⓿ Either it is seen to be failing,

⓿ Or it is no longer necessary.

It is essential that money is not wasted on ineffective Risk Plans or the value of the entire process may be questioned.

10.9.7.8 Closing a Risk Plan

It is always best to err on the side of caution when closing a Risk Plan i.e. if in doubt, leave it open. When the Risk Plan is thought to have met its objectives:

⑩ Obtain agreement from the Risk Owner to close the plan,

⑩ Trust a statement or promise only when it is fully endorsed and recorded at the right level in the enterprise,

⑩ Ensure that all the necessary changes resulting from the Risk Plan are incorporated into the main project plans, documentation and specifications,

⑩ Check that any revised practices are reflected in the business area,

⑩ Evaluate if there is any remaining (residual) risk which may still need to be addressed.

Then document the closure in the Risk Administration System.

10.9.7.9 Configuration Control

To allow the Project Management process to be effective, the main projects and management concerned must have an effective change and configuration control system in place. To support the Risk Management element, some form of Risk Administration System will be required, in either a manual format for small projects, or a software based tool (such as Risk) for larger projects.

10.10 Applying The Sigma Process - Practical Considerations

10.10.1 Project Prioritisation Considerations

It is important that the project inventory is updated regularly. Ensure it relies on high quality information, otherwise the risk management exercise will quickly become meaningless (i.e. projects currently being assessed for risks may meet their objectives whilst a new, more critical project flounders because it has not been identified for risk assessment.)

It is, also, important that all relevant contacts are involved in the process (i.e. technical and business). The projects should be prioritised from the overall business perspective, not that of a single department.

10.10.1.1 Using software tools to maintain the project inventory

A sophisticated system is not always necessary for maintaining the project inventory. A simple Word document or spreadsheet can do the job adequately. The main questions to ask when deciding how to record project information, including data for the Criticality/Complexity matrices, are:

⑩ What is the inventory going to be used for?

> The inventory is useful for deciding on resource allocation, as well as which projects are approved to go ahead and which should be assessed for risk. The inventory may need to be sorted by various criteria, for example by business criticality, start date, department, even by project manager. If the inventory is going to be used in this complex manner, it may be worth using Microsoft Project or Access database, or something similar.

⑩ How large is the inventory going to be?

A documented list of projects and their attributes has been used for an inventory of up to 50 projects, without recourse to more complex software.

⑩ Is there a system for prioritising projects already in place at the client? Can this easily be adapted?

Some clients have been found to have sophisticated yet disparate and unconnected lists of projects, prioritised through departmental influence and politics, rather than by a sensible business level view.

Overall, the purpose of the project prioritisation process is to ensure that there is a sensible, business-level view of which projects contribute most to the business strategy and which need a formal risk management process to increase their chance of successfully meeting their objectives.

Further value can be added through improving resource allocation and the client's general management of projects.

It is not always necessary to visually represent projects using the CC Diagram. Some clients prefer to prioritise projects based solely on the business criticality, and use the other parameters for deciding which projects require formal risk management.

Even then, other management systems can lead senior management to highlight particular projects for formal risk management. For example, through their own discussions with the project managers. Even in these circumstances, plotting the projects on a CC Diagram can help to focus priorities and ensure consistency across the business.

10.10.1.2 Assessment Analysis Considerations

It is important to identify the right people to interview in order to capture assessments. You will need to interview people from the project organisation and the customer/users departments.

These people tend to be:

⑩ Project managers (smaller projects):

If the project organisation is relatively small, the project manager is likely to be a rich source of information. The project manager should have a good understanding of the project plans and the activities taking place within the project. He should be able to speak on behalf of his direct reports on the project.

⑩ Project managers (larger projects):

In larger projects, the Project Manager will tend to identify his Team Leaders for interview. The Project Manager will be able to provide assessmens relating to the project as a whole and the events it faces relative to overall business objectives whilst the Team Leaders will tend to concentrate on their own area.

A project manager looking at assessments raised by the project team will be able to provide feedback and further assessments.

⑩ Team members:

Down to a certain level, team members should be interviewed to capture a assessments. They have the front line view and more detailed knowledge of the events and potential problems they could face in delivering the project.

Clearly it is not necessary to interview everyone including the student on work experience. There is a pragmatic cut-off point where you have to rely on the communication mechanisms within the team ensuring the more senior members have a good idea of what is going on.

Having said that, if you are getting assessments that are clearly lacking in detail and specifics, it may indicate that you do need to interview more people further down the project organisation.

◎ Customers/Users:

Interview those who would suffer most from the failure of the project to deliver, and those who will need to provide resources to the project in the form of finance or future users

10.10.1.3 Getting To The Right Assessment

The Sigma process is designed to capture and analyse a substantially complete set of critical project assessments.

To achieve this the following criteria must be met:

◎ The key contacts in the organisation are interviewed to capture the assessments,

◎ The interviews are conducted by a team trained in the Sigmqa methodology,

◎ Someone with experience of the business and project management is involved, to help capture assessments that are not implicit from the plans,

◎ The assessments captured are disseminated around the key players to gather counter/multiple viewpoints on the ratings,

◎ The key players are re-interviewed on a regular basis to review the assessments and to capture new assessments.

In practice, interviewing is the most efficient and successful method of capturing high quality assumptions. Other methods have been tried. However, it is far more difficult to ensure that high quality data is captured.

For example:

◎ **Self-analysis**:

Every Key Player needs to be fully conversant in Sigma and its underlying principles, be knowledgeable about project management, and apply a degree of objectivity to their own thinking.

Furthermore, people tend to need prompting to perform this exercise, even if they think it's a great idea, a separate facilitator ensuring regular review would be required. Great caution is required with this approach if quality problems are not to negate all benefits.

◎ **Forms**:

A structured form, which allows project team members to fill in their assessments and rate them is clearly a form of self-analysis and, in practice, a very ineffective means of identifying risks. The form will be seen as an administrative burden and will not receive the attention it deserves.

The information will inevitably be brief incomplete and unclear and the ratings will be meaningless without the challenge of a facilitator to ensure that the assessment is the right one and the ratings are fully understood and allocated relative to the project and business objectives.

⑩ **Workshops**:

If led by a Sigma expert, workshops can produce a lot of data in a short space of time. The data quality however tends to be low, unless significant extra time is spent improving it. People may become more reticent in divulging potentially sensitive information, and assessments tend to be generic, embodying the group point of view and losing much of their specific value.

One effective way to run a risk assessment workshop is to use a Project Readiness Assessment Walkthrough.

Compared with other approaches, the primary benefits of the interviewing approach are:

⑩ Only a small number of interviewers need to be Sigma experts and have experience of project management to generate high quality output,

⑩ Key contacts do not need to be highly trained in Sigma. They only need to understand how information is graded and how the process works (at a high level),

⑩ Key contacts will talk more openly and provide higher quality information in a one-on-one situation, and differences of opinion are made visible to the interviewer,

⑩ The interviewer is not a stakeholder and can be completely objective,

⑩ The interviewer can judge that the right contacts are regularly interviewed to maintain a complete assessments register.

10.10.1.4 Preparation For The Interview

Key contacts are interviewed not only to capture their assessments and risks, but also to communicate relevant information captured from other key contacts. This can actually lead to further assessments and risks being captured as the interviewee reacts to the new information. Thus, a decision has to be made as to what other assessments should be shown to the Key Player being interviewed.

Ideally each key contact should be aware of assessments and risks raised by all key contacts for the project, and would benefit from the latest report of these on each visit. Should the project prove too large for this to be practical, assessments and risks can be categorised and a sub-group of these reported to each key contact. The risks database facilitates this process.

10.10.1.5 During The Interview

If this is the first time the interviewer has seen the person, they should talk through the Assessment Analysis approach, and determine the correct role of this person in the process.

It is also important for the interviewer to understand the person's role in each project they are a key contact for, so that questions can be directed effectively.

There are a number of key activities that should take place during the interview:

⑩ Check that plans exist for the project and at least for the current phase. Work from a high level milestone plan if it is available or the plan that they are carrying around in their head at the very least.

⑩ Walk them through the plan asking the question "what is your next milestone and do you have any concerns in achieving it?"

⑩ Ask questions based on what planned activities need to take place, focusing on weak areas, activities that are not in the plans and suggest potential problems the interviewee may not have thought of.

⑩ Capture important assessments that have been made explicitly in the plans, or that the interviewee has revealed as a result of your questions.

⑩ Should an assessment be graded as C,C or higher it will be necessary to capture the risk statement straight away. Ask the interviewee what the consequences will be if the assessment does not happen, and capture all other details required to describe the risk.

⑩ Existing assessments and risks should be reviewed with the interviewee. This may lead to a change in grading, wording, timing, or any other attributes of an assumption or risk. The assessment and risk wording and ratings must be the Key Player's and not the interviewer's if ownership and buy-in are to be encouraged.

10.10.1.6 At The End Of The Interview

There are a number of steps to take in completing an interview:

⑩ Review the assessments and risks to ensure the interviewee is happy with the wording and ratings,

⑩ Make sure they represent what they believe to be their major risks/concerns,

⑩ Explain that you will revisit these assessments the next time you visit, and that meantime their risks will be reviewed and action taken by senior management.

10.10.1.7 Review Assessments And Risks Regularly

It is important to review assessments and risks on a regular basis because:

⑩ There should be a constant communication of this information between key contacts and senior management,

⑩ The quality of the assessments changes over time; they may become more or less stable as further information is made available, or the project may become more or less sensitive to it as planning decisions are made,

⑩ Action to control risks and assessments needs to be communicated back to the originators and any resulting changes to the list made according to the originator's feedback.

There are a number of tools that can be used to manage this:

⓿ The Assessments and Risks Register,

⓿ A matrix of key contacts against projects being tracked for risk,

⓿ An interview schedule for all key contacts,

⓿ An interviewing template to provide structure during the interview.

These must be maintained on a regular basis in order that the risk management

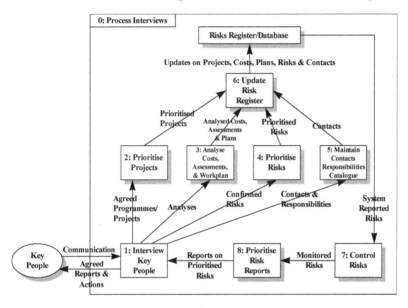

The Interviewing Process

process remains focused.

11. Transferring Ownership Of The Sigma Process

11.1 Sponsorship And Ownership

The Sigma process can only be effectively handed over to the client organisation, or users when two primary criteria are met:

⓿ A clear sponsor (champion) and new process owners are identified,

⓿ The process is fully up and running and is stable.

11.1.1 Identifying The Process Champions

It is important to ensure that the future sponsor and key people who will be using the process have been involved in it and can testify to the value that it adds. It may still be necessary to continue selling the benefits of the process throughout the hand over period.

The choice of sponsor is critical, since they will have to be able to keep the process on track, reinforcing it where necessary and ensuring it remains focused on what is important to the business.

Their main responsibilities would be:

⑩ Enable the necessary interviews and break down any local resistance,

⑩ Ensuring the senior management risk review process is kept focused, action-oriented and timely,

⑩ Managing any conflict between senior management in the process,

⑩ Applying pressure on those who try to subvert or hijack the process..

11.1.2 Identifying The Process Owners

Equally important as the choice of sponsor is the choice of client process owners. No matter how strongly the sponsor champions the process, ineffective process owners can allow the quality of information from the Sigma method to slip, rendering the functions almost valueless in a matter of weeks following hand over.

There are several critical success factors for ensuring the effective management of the Sigma process:

⑩ Affording time and resources for the process,

⑩ Enthusiasm and attention to detail,

⑩ Access to key players in the organisation.

11.1.2.1 Affording Time For The Process

The amount of effort associated with the process will be far greater than the new process owner will expect.

There may be a tendency to treat the process as a low priority activity to be done in the process owner's spare time. If care is not taken the process will not receive the necessary attention and will quickly fall down the priority list when compared with other direct business responsibilities.

Sometimes the new process owner will see this new responsibility as a poor career move. Under these circumstances there are a number of benefits that should be emphasised:

⑩ Personal exposure to senior management on a regular (trusting) basis,

⑩ Access to all relevant areas of the organisation, affording a much improved cross-functional view of the organisation and a larger network of contacts,

⑩ Greater understanding of the projects being undertaken and the business problems they face,

⑩ Greater appreciation of project management principles and the business as a whole.

11.1.2.2 Enthusiasm And Attention To Detail

An essential ingredient of a successful implementation of the Sigma process is high quality data. The process owner should be able to demonstrate a meticulous attention to detail. If senior management cannot understand the risks or the risks are not presented to them in a consistent manner, they will quickly lose interest and question the value of the process.

Typical groups within an organisation which could provide suitable process owners are:

⑩ Internal consulting groups,

⑩ Programme Offices,

⑩ Quality management groups,

⑩ Audit departments,

⑩ Organisation-wide project/programme management groups.

However, care must be taken with how positively these groups are perceived within each organisation.

11.1.2.3 Access To Key Players In The Organisation

The process owners have to be in such a position as to be able to access:

⑩ Key Players: project sponsors, project managers and key project contacts,

⑩ Senior Management contacts: Sigma process sponsor, senior managers involved in risk review and control,

The process champion will need to intervene quickly in any access problems.

11.1.2.4 Avoiding The Split-up Approach

There may be a temptation to split up the process across the organisation to spread the load and allow part time involvement of the new process owners. Great care is required here as one the key benefits of Sigma, that of providing a cross-organisational communication mechanism, may be undermined.

11.2 Hand Over Process

11.2.1 Training

The training must cover both the theory and practical application of Sigma. Theory without practice inevitably leads to poor quality data and practice without theory extends the learning curve considerably.

11.2.2 Parallel Interviews

One of the most important parts of the hand over programme is the parallel interview sessions. After the training course, a series of parallel interviews are held where the consultant and new process owner conduct a number of interviews together.

This achieves the following:

⑩ The new process owner can learn first hand how the interviews are conducted and how the different interviewees behave,

⑩ The new process owner can practice the interviewing process with a degree of comfort that they can make mistakes,

⑩ The interviewees can visibly see the handover process at work and meet their future interviewers.

Past experience has shown that approximately 15-20 parallel interviews are required to get the average candidate up to speed.

It is useful to arrange for the new process owner to conduct an interview with each of the Key Players during the parallel interviewing process. This allows them to establish a dialogue with each of their contacts by the time the hand over is complete.

It is also useful to interview some new Key Players, so the new process owner can learn how to explain the process and its benefits to the individual.

Parallel interviews should be conducted with the consultant leading for the first few interviews, and then with the new process owner leading in later interviews. Prior to each interview, the consultant should review with the client the documentation required for the interview. This avoids the client losing valuable learning time in the interview being unable to understand the language used or the relevant project background.

11.2.2.1 Consultant Leading Interviews

The main skill learnt by the client through this activity is how to capture high quality assumptions and risks.

The new process owner should have a copy of all the materials that the consultant has to conduct the interview. The consultant conducts the interview as normal. The new process owner will learn how interviews are conducted, and practice capturing assumptions and risks as the consultant asks the questions. When the captured data is recorded in a risk register, the consultant can correct the client's notes after the interview. About 5 to 10 interviews should be sufficient.

11.2.2.2 Client Leading Interviews

One of the most difficult aspects of the interviewing process is asking the right questions and maintaining a dialogue with the interviewee whilst at the same time following a strict code in documenting the assessments and risks and their associated grading. The client will have acquired skills in the mechanics of capturing assessments and risks, and will now have the opportunity to practice asking the right questions to obtain that information.

The consultant should allow the client to run the interview by themselves as much as possible, intervening only where quality is compromised to ensure the best results from the interview are obtained. The consultant should concentrate on asking any questions at the end that may have been missed due to lack of experience.

The consultant should spend time with the new process owner after the interview to discuss what went well (so the client knows they are progressing) and what could be improved.

More time should be spent on this stage of the hand over with the client leading the interviews since they have more to learn at this stage. About 10 to 15 interviews should be sufficient.

11.2.3 Procedures And Supporting Documentation

The client process owners have a lot to learn in a short hand over time i.e.: a complete methodology, the current process, the interviewing technique and using the software support tools. All this needs to be supported by some form of procedures, to provide a reference point during the hand over and to act as supplementary training material should the process owners need to train further people in their organisation.

11.2.4 Software Tool Support

Not only do the new process owners have to learn to use the risks software, but someone will have to be trained in maintaining the software, as it is likely at some stage that they will need to add new fields or categories, or modify the report formats.

The ideal way to achieve this is to provide good documentation of the system, combined with at least half a day's worth hands-on demonstration with the chosen system maintainer. This person should be able to 'play' with the database by themselves (trying to add fields/modify reports etc.) and be able to come back and ask questions before the hand over is complete.

11.3 Ongoing Quality Management

11.3.1 Can they Continue To Do It Themselves?

No matter how well the hand over has been managed the process can still start to fail quickly when the consultants depart if the new process owners do not fully understand all the subtleties of the application of Sigma. A way of minimising this risk is to consciously involve the new process owner in all decisions, discussions, changes etc. during the hand over period.

When adjustments need to be made to the process, involve the new process owners - ask how they would deal with it first. Such adjustments may include:

- Reassessing the list of key contacts where necessary,
- Changing the format of risk reporting to senior management,
- Adjusting interviewing approach or timing,
- Changing the format of the risk review meetings,
- Managing political hiccups that can disrupt the process,
- Seizing opportunities that can enhance the process,
- Provide them with a 'troubleshooting guide' which highlights typical symptoms, what it is that may be going wrong, and ways to correct the problem.

  For example:

  - Not identifying many risks,
  - Senior management do not understand the risks,
  - Key Players start to lose interest in the process,
  - The Risk Review meetings become irregular.

11.3.2 Applying Sigma To The Hand Over Phase

A highly effective way of ensuring a successful hand over within agreed time scales is to apply the Sigma process to the hand over activities. This means treating the hand over as a project and forming a simple Risk Register which is communicated to the client at regular intervals.

If the Process Sponsor is regularly updated with risks to achieving successful hand over, he will be able to take actions to support the consultants. Furthermore, if the hand over is impeded by factors outside the consultants' control, for example late sign-off, problems accessing the right people, and so on, the risk register can provide a valuable audit trail backing up promises to deliver, or supporting any explanations as to why there were delays.

11.4 Integrating Methodologies

In handing over and in applying any Risk Management method, as in any other Management tool, the integration of all existing methodologies is of primary importance.

In choosing the new owner of the Sigma methodology, consideration must be given regarding the experience possessed by the new proprietor. Preference should be given to the people with systems methodology and management procedures training and experience.

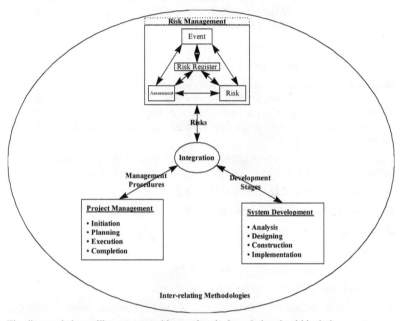

The diagram below will serve as a guide on what the knowledge should include.

12. Interfacing

12.1 The Diagrammatic Representation Of The Desired Integration

The diagrammatic flow shown below is the desired overall integration of related top-level processes:

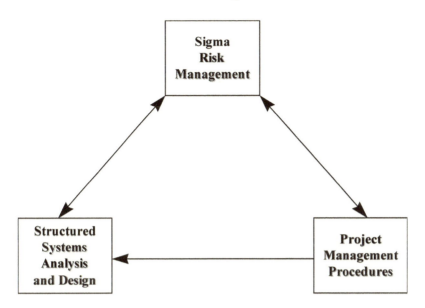

Integration Of Methodologies

12.2 Suggested Interfacing

The suggested interfacing of the Structured Systems Analysis and Design methodology and the Project Management procedures to the Risk Management processes may be done separately as shown in the diagram below:

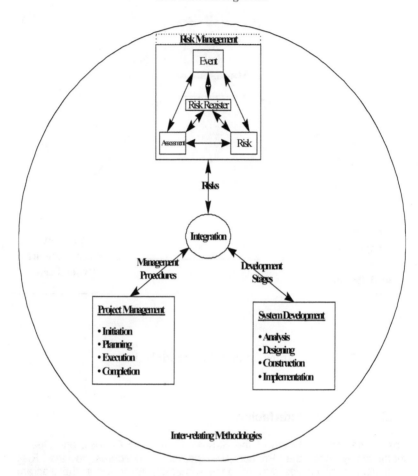

Inter-relating Methodologies

12.3 Using The Existing Methods

In interfacing the various processes from the different existing organisational methodologies, the main steps of Project Management and the Structures Systems Analysis and Designing methodology, will be incorporated in this exercise.
Namely, the six main stages of system development:

- ⑩ Analyse,
- ⑩ Design,
- ⑩ Construct,
- ⑩ Implement.

The identified processes of Project Management procedures and Structured Systems Analysis and Designing to be linked to the Sigma Risk Management processes of:

- Events,
- Assessments,
- Risks,
- Plans.

The actual points of interfacing the stages of development, the procedures of managing projects and the assessments analysis of the Sigma methodology, are explored in this book.

12.4 Connection Of Project Management To Sigma

Regarding the connections of the project management procedures to risk management, it is suggested that the link be from the higher level of project management work elements, such as:

- Initiate Risk Management,

- Adjust Project Approach to Risk,

- Control Risk,

- Complete Risk Management.

This is as far as most users drill into the process and generally as far as they are required to by structured analysis and designing methodologies guidelines.

13. Managing The Sigma Risk Programme

13.1 The Diagrammatic Representation Shown.

⓪ O: Sigma Risk Management Decomposition,
⓪ O: Management of the Risk Programme,
⓪ 1: Managing of the Sigma Process,
⓪ 2: Facilitation of the Sigma Programme,
⓪ 3. Practising the Sigma Methodology.

13.2 O: Sigma Risk Management Decomposition

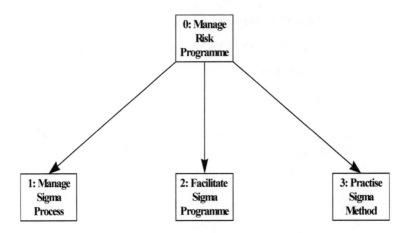

The Sigma Risk Management Decomposition

13.3 O: Management of the Risk Programme

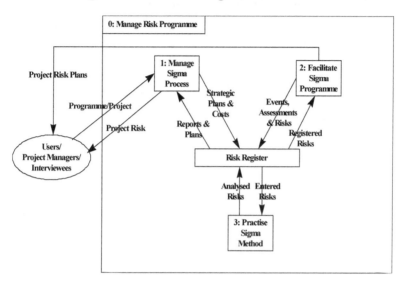

13.4 1: Managing of the Sigma Process

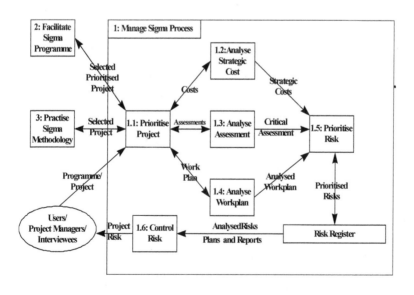

13.5 2: Facilitation of the Sigma Programme

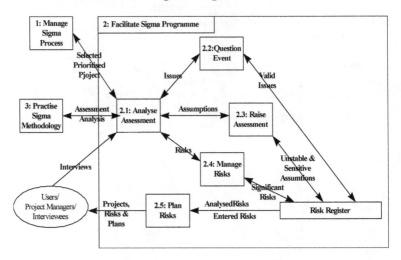

13.6 3: Practising the Sigma Methodology

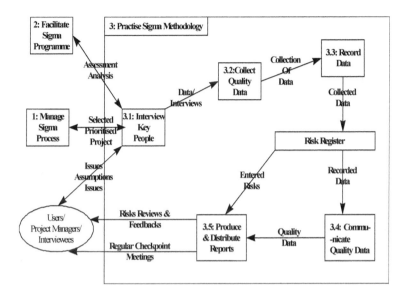

13.7 Relationship Of Dataflow Diagrams and SIGMA

All dataflow diagrams shown in above pages are based on the PsySys Limited manuals and handouts as used for the Sigma training.

13.7.1 DATAFLOW DIAGRAMS: NUMBERS AND TITLES.	DESCRIPTION OF DIAGRAMS AND THEIR PROCESSES.
0: Manage Risk Programme	Project: something complex that you want (plan) to happen. Risk: Something that you don't want to happen. Project management: Planning and making things happen. Risk management: Attacking anything that might disturb the plans
1: Manage Sigma Process	The Sigma process consists of an integrated closed loop method which logically progresses through: I1 Project Prioritisation, I2 Strategic Cost Analysis,

I.T. Risk Management

	I3 Assessment Analysis,
	I4 Work Plan Analysis,
	I5 Risk Prioritisation,
	I6 Risk Control.
1.1 Prioritise Project	Only complex and critical projects need to have a fully structured risk management process in place.
I1 Analyse Strategic Cost	Strategic Cost Analysis provides a means of assessing the cost risk in a project in its very early stages and "kick-starts" the risk management process.
1.3 Analyse Assessment	Fundamentally, projects only fail due to two reasons either the wrong assumptions were made or the significance of the assumptions was not understood.
I1 Analyse Work plan	Work Plan Analysis may be used to focus on the risky areas of detailed, multi-level project plans when time is of the essence.
1.5 Prioritise Risk	Prioritisation allows the Project Manager to divert limited resources at the most critical project risk.
1.6 Control Risk	Risks may be attacked at both the strategic and tactical levels. Strategic approaches look for trends and underlying causes for groups of risks. Tactical approaches take each risk at face value.
2: Facilitate Sigma Programme	The Sigma Programme is essentially a framework process that allows the capture of collective knowledge and viewpoints from those involved on the project, in a form that facilitates communication of events, assessments and ensures pro-active management of risks. By dramatically improving communication, risks are avoided, or managed proactively and project objectives are delivered on time.
2.1: Analyse Assessment	The core of Sigma is Assessment Analysis. This uses structured techniques to analyse project plans and identify the most sensitive assumptions that are potentially unstable, and therefore the source of greatest risk.

I.T. Risk Management

2.2: Question Event	Events are open questions which are holding up plans/implementation. An Event is any open question which has been asked at the right time to which a high quality answer cannot be provided without escalation.
2.3: Raise Assessment	Many Events are closed by making Assessments in plans. An Assessment is a single, simple, positive or negative statement.
2.4: Manage Risks	Unstable/sensitive assumptions create risks. Significant risks need to be managed formally. Definition: A Risk is a simple statement of the form: "IF" Assumption proves incorrect, "THEN" Describe the impact.
2.5: Plan Risks	Risk Plans impact project plans. Events, Assessments and risks are inherent in the project plans. Population of assessment and risks registers by progressing Risk Plans/Main/Project Plans.
3: Practise Sigma Methodology	The Role of a Sigma Risk Management Practitioner includes: Interview 'Key People' within the project, To collect 'Quality' data, Ensure the data collected is recorded in the Risk Register, Communicate the 'Quality' data to Project staff, Produces accurate and timely reports for meetings: Weekly Check Point Meetings, Risk Review Boards.
3.1: Interview Key People	Identifying the right people to interview is critical to producing a comprehensive and coherent picture of the risks facing a project. So, to decide who should be interviewed, start with the project or programme organisational structure.

I.T. Risk Management

	Depending on the scope of the risk assessment it may be necessary to map the organisational hierarchy to ensure that the right people are interviewed and that the risks arising are reviewed at an appropriate level.
3.2: Collect Quality Data	This function requires to: Interview 'Key People' within the project, Collect 'Quality' data.
3.3: Record Data	Having Interviewed the 'Key People' within the project and collected the 'Quality' data: Ensure the data collected is recorded in the Risk Register.
3.4: Communicate Quality Data	The interviews of 'Key People' within the project have been completed, the 'Quality' data recorded in the Risk Register and communicated to Project staff, the next function is: Facilitate and ensure the Risk Management process stays on track.
3.5: Produce Reports	After the initial round of interviews, a suitable forum must be established to discuss the risks identified. The best method is to establish a specific Risk Review Meeting with a representation consisting of the Risk Owners and chaired by the Programme Director or the process 'champion' in the client organisation. The main function is to: Produce accurate and timely reports for meetings: Weekly Check Point Meetings, Risk Review Boards.

Detester/Database Of Events, Assessments and Risks Register Description.

Data Store:	All events, assessments and risks captured should be held in a Risks Register.
Risk Register	Remember, only critical assessments will be converted into risks and held in a Risk Register.

I.T. Risk Management

Thus, by filtering the assessments and consolidating them into risks, all information captured will be rationalised and details of their source and consequences will be traceable.

14 Project Management Improvement

14.1 Weakness In Project Management

In the process of using Sigma, it will soon become apparent that probably the main cause for the threatening risks is project management, or perhaps the weakness of individual/s to manage the growth of a system.

Even if you used Sigma in its fullest, your experience will soon enable you to identify the real cause for the failure in implementing your system. In such cases, your programme/project management may need further support, assistance, training, better communicational ability and proper delegating, or even some listening to other people involved in the on-going project. Employing the expertise of a consultancy may help.

The main point is to try to reduce potential project caused loss by providing efficient *Event* driven project reviews for the critical project/s. Such project consulting steps will create and utilise virtual group of experienced project managers. As a panel of experts they will assess critical projects and provide consulting being perceived as helpful by the project team, the management and the users/clients.

14.2 Process Steps Required

For such a project consulting programme implementation, various process steps are needed:

1. Select critical project,

2. Understand project status,

3. Plan project review,

4. Create reviews agenda

5. Execute project review,

6. Implement change plan,

7. Conclude the project review.

14.3 Collection Of Information

The project manager of the project pending the review should gather the requested information from the existing project documentation. If such documentation is not available, the project manager should collect as much a as possible from verbal discussions.

Such a preparation should include details of:

1. Communication plans,

2. Project organisation plan,

3. Contacts and scopes,

4. Background and status of finances,

5. Schedules,

6. Status and history of resources plans,

7. Quality plan,

8. System documentation,

9. Brief description of the project environment,

10.Risk register prioritisation/s and other reports,

11.Last project review minutes.

I1 Agreement On Resources

The final outcome of all communications, discussions and reviews should enable the team members to produce an agreed score on the various project management resources.

The list for such a scoring agreement should include:

1. Quality management,

2. User participation,

3. Requirements management,

4. Communications,

5. Business orientation,

6. Project team,

7. Project planning,

8. Risk management,

9. Technical environment.

14.5 The Score Graph

A scoring card graph can easily be produced on a spreadsheet and it ought to look something like this:

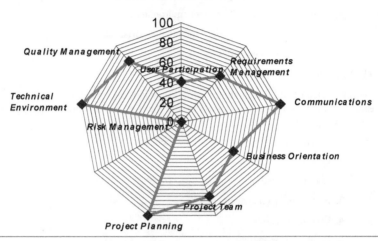

Score Card Graph.

15. Remembering The Main Risk Management Points

A risk is an uncertain event which may have an adverse effect on the project's objectives. Sigma is a proven risk management methodology, which should be very effective in the quest for identifying risks throughout the project life-cycle.

Remember, risk management is:

1. Forward looking, investigating problems and how to deal with threats,

2. A tool enabling communication, getting people at all levels to talk to each other and to interact,

3. A no blame team culture, bringing concerns into the open where actions can be taken and plans put in place, in order to stop a risk occurring.

The Sigma process commences by identifying the enterprises most important and risky projects, as these must be given priority. Sigma is essentially a method that permits the collection of knowledge and experience from those involved, in a form that facilitates the Systematic Interaction and Generic Methodology for Applications:

Σystematic: The varied events, their assessments and the consequential risks relating to or consisting of a system. Methodical in procedures and plans, these are addressed to those involved and deliberating within the parameters of their systems development responsibilities. The results being dependable on:

Interaction: The mutual or reciprocal action which encourages those involved in the programmes and projects to communicate with each other and to work closely with a view to solving the threatening events before they impact on the development of the system. The individuals involved maintain a -

I.T. Risk Management

Generic approach, which relates and characterises the whole group of those involved in assessing the events and attacking the threatening ones before they become risks to the development of the system. The end result being the avoidance of apparent problems within the pre-defined users systems requirements. This is enabled by following the -

Methodology: The system architects and the risk management practitioners simply follow the approved body of systems development methods, rules and management procedures employed by their organisation. For practical or even ethical reasons, it must be noted that with such a philosophy, it is seldom possible to fulfil all requirements of very large organisational systems. As such, the *Sigma* methodology is administered in -

Applications: Putting to use such techniques and in applying the risk management principles in the development of various *applications* will involve numerous and varied activities. A concrete issue in developing new applications is the problem of communication among the people involved, the motivation constantly needed for *generic* work, the ability to *interact systematically* and in using a structured systems *methodology*.

INDEX

Acceptable   31

Access Control And Maintenance 20

Access To Key Players In The Organisation 75

Action Managers   65

Actions   6

Administration System Tool   32

Analyse 80

Analysis 18

Analysis And Quantification of Risks   35

Approaches to Risk Management   34

Assessment   70, 80

Assessment Analysis   31, 50

Assessment Analysis Considerations   70

Assessment Of Risks 18

Assessment Team 45

Assessments and Risks Register   56

Brainstorming   35

Business Case   26

Business Complexity 47

Business Continuity 23

Business Related Overview 23

Business Responsibilities   23

Categorisation Of Assessments — Risk Drivers   52

Collection Of Information   90

Communicating The Risks 32

Communication   18, 30

Communication Of Assessments   37

Concept 11

confidence levels (A-C)   55

Configuration Control   68

Construct 80

Content Security 22

Control   30

Control and Lack of Follow-Through   35

Controllability 56

Convert Key Assessments Into Risks 51

Corporate Governance   23

Corporate Reputation   24

Criticality 56

I.T. Risk Management

Cryptography 21

Customers/Users 70

Cycle   12, 18, 30

Dataflow Diagrams and SIGMA   85

Decision 59

Decomposition   82

Definition of a Risk   33

Design 80

Developing Risk Plans   62

Diagrammatic Representation Of The Desired Integration 79

Diagrammatic Representation Shown.   82

Do It Themselves 77

Enthusiasm And Attention To Detail   75

Events 80

Events And Risk Registers   34

Facilitation of the Sigma Programme   84

Features And Benefits Of The Sigma Approach   30

Firewalls 20

Flexible   30

Forms 70

Global Networked Business   26

Hand Over Phase   78

Hand Over Process   75

Hot Spots in the Plans   54

Identification and Analysis   49

Identify the Sources of Assessments   50

If (incorrect), Then (consequences) 51

Impact On Business 15

Implement 80

Implementation of a Risk Management Strategy   27

Information   30

Information Assurance   24

Information Availability   25

Initiating the Sigma Process 43

Integrating Methodologies 13, 78

Interdependency And The Critical Information Infrastructure 25

Inter-dependency On I.T.16

Interfacing   79

Internal Control 16

Internet Commerce   25

Inter-relationship 13

Interview   34, 44, 72, 76

Intrusion Detection Systems (IDS) 21

Management of the Risk Programme   83

I.T. Risk Management

Managing of the Sigma Process   83

Managing The Sigma Risk Programme   82

Matrix updated   56

Meeting 66

Methodology Explained   28

Methods   80

Milestone 52

Milestone 59

Monitoring   27

Obligations Of A Corporation 16

Organisation And Responsibilities   65

Organisational Culture 43

Organisational Structure and Stability 43

participants   55

Plan Is The Baseline   38

Plans 61, 62, 63 67, 68, 80

Plans for Attacking Assessment Based Risks   62

Plans for Attacking Planning-Type Risks   6

Plans to Attack Risks   62

Policy 52

Positioning and Prioritising New Projects 48

Positioning in time

Positioning Risks   56

Practical Considerations   69

Practice 18

Practising the Sigma Methodology 84

Preparing For A Crisis   24

Pre-selected Assessments   56

Principles Of Risk Management   33

Principles of the Sigma Methodology   37

Prioritisation   40, 47, 48, 50, 69

Procedures And Supporting Documentation   77

Process 37, 74

Process Champions   74

 Process Owners   74

Process Overview   39

Process Steps Required   90

Programme Management   11

Programme Objectives   28

Programme15

Project And/Or Programme   47

Project Approval and Resources   48

Project Fully Planned And Proceeding   44

Project In Trouble   44

Project Inclusion In Process   47

Project Management   90

I.T. Risk Management

Project Management Improvement   90

Project Management, Sigma   81

Project managers (larger projects) 70

Project managers (smaller projects) 70

Project Size 47
Project Status   44

Projects Change Position on the CC Diagram   49

Projects Formal Risk Management 49

Quality Management   77

Quality Review of Plans   54

Quality Using the Sigma Scale   38
Reference Matrix   48
Representation 66
Reputation Management   23
Resources 59, 91
Review Assessments And Risks Regularly   73
Review Board   66
reviewed regularly   56

Risk Control   41, 59

Risk Management Points   92

Risk Management vs. Project Management   36

Risk Owners   65

Risk Ownership and Risk Action Managers   65

Risk Prioritisation 40, 56

Risk Register Reports   58

Risk Review Meetings   46
Risks 80

Safeguarding Information Systems 20

Score Graph   92

Self-analysis 70

Sensitivity 50

Sigma Methodology   11, 28

Software Control And Maintenance 20

Software Tool Support   77

software tools to maintain the project inventory 69

Split-up Approach   75

Sponsorship And Ownership   74

Stability 50

Strategic Approaches   59

Strategic Cost Analysis   31, 53

structure.   55
Tactical Approaches   61
Team Approach   33
Team members 70
Technical 52, 59
Technical Complexity 47

I.T. Risk Management

Timing 56

Training 17, 75

Transfer 36

Transferring Ownership Of The Sigma Process 74

Uncertainty Equals Risk 38

Use The Contents Of This Book 32

Virus Protection 22

Vulnerability Assessment 21

Walkthrough 54

Walkthrough is completed 56

Walkthrough is scheduled 5

Walkthrough starts by concentrating on the Cs 55

Work Plan Analysis 32, 53

Workshops 70

World Wide Web (WWW) 26

I.T. Risk Management

BOOK PUBLICATIONS: BY EURING PROF DR ANDREAS SOFRONIOU

1. TITLE: THERAPEUTIC PHILOSOPHY FOR THE INDIVIDUAL AND THE STATE
ISBN: 0 9527956 5 5
DESCRIPTION: Concepts and didactics of philosophers through the ages. From the Hellenic rhetorics, to recent European schools of ideas. The logic of Therapeutic Philosophy expressed assists in the understanding of human behaviour and the way in which philosophy can contribute to the treatment of the individual and the state.

2. TITLE: Philosophic Counselling For People And Their Governments
ISBN: 0 9527956 6 3
DESCRIPTION: The logic of philosophic counselling assists in the understanding of human behaviour and can contribute to the treatment of the people and their governments. Philosophic Counselling has been with us since the times Pericles, the Golden Age of Athens. As such, this book includes the ideas of Plato, Aristotle, Machiavelli, Hobbes, Locke, Rousseau, Hume, Burke, Hegel, Bentham, Mill, Marx and other philosophers in contemporary counselling.

3. TITLE: A TOWN CALLED MORPHOU.
ISBN: 0 9527956 2 0
DESCRIPTION: The forceful simplicity of the majority of these verses own their existence to the adoration of the opposite sex, freedom, patriotism, the didactics of Aristotle, Freud, Comparative Religion, and the belief in family unity.

4. TITLE: EXPERIENCE MY BEFRIENDED IDEAL
ISBN: 0 9527253 0 4
DESCRIPTION: This anthology consists of metaphysical poems, verses with philosophical simplicity and romantic compositions.

5. TITLE: Joyful Parenting.
ISBN: 0 9527956 1 2
DESCRIPTION The psychology of child culture. Pre-natal, post-natal and all stages of development.

6. TITLE: The Management Of Commercial Computing.
ISBN: 0 9527956 0 4
DESCRIPTION The development and management of systems and people in multi-national corporations, systems and software houses, government departments, European Union Commissions and academia.

7. TITLE: Business Information Systems, Concepts and Examples.
ISBN: 0 9527956 3 9
DESCRIPTION This book aims to fill a gap in the current business and tutorial literature. It has been designed for the business individual, for the student and the computer professional who need a detailed overview of business information systems. It explores computing in general, the structured development of systems using processes and data analysis; object oriented and other methods. It includes the project planning and testing procedures for the Millennium thread.

8. TITLE: A Guide To Information Technology
ISBN: 0 9527956 4 7
DESCRIPTION The book covers the fundamental aspects of computing and the development of new information systems. Explains the current systems, structured analysis and designing, management, planning and the year 2000 problems and solutions.

9. TITLE: Trading On The Internet In The Year 2000 And Beyond
ISBN: 0 9527956 7 1
DESCRIPTION: Use of the Internet and E-Commerce is a business issue first and foremost. The Information Superhighway will see the consumer having access to a myriad of data through the PC or TV screen. The digital market is so extensive that most retailers will establish the marketplace by designing around a number of architectural models. The design of the system will be based on how the users work and what suits the overall business environment.

10. TITLE: The Sigma Methodology For Risk Management In Systems Development
ISBN: 0 9527956 8 X
DESCRIPTION: The Sigma methodology allows the capture of collective knowledge and expertise from those involved on the project, in a form that facilitates communication of Events, Assessments and the pro-active management of Risks. Sigma can be applied to any type of project, or programme.

I.T.
RISK
MANAGEMENT

The I.T. Risk Management methodology allows the capture of collective knowledge and expertise from those involved on the project, in a form that facilitates the communication of Events, Assessments and the pro-active management of Risks. Sigma can be applied to any type of project, or programme. In essence, this is the mechanism by which the functions of programmes and projects are held together as a result of the principles operating within the *Sigma* methodology.

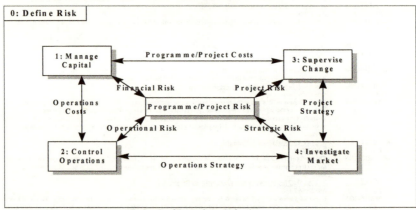

EurIng Prof Dr Andreas Sofroniou

Born in Morphou, Cyprus, Andreas holds the degrees of Doctor of Psychology and Doctor of Philosophy. Also, qualifications as a chartered member and fellow of eighteen British professional institutions, including Engineering, Systems, Computing, Directing, Complementary Medicine, Management, Production, Programming, Marketing, Petroleum, Data Processing, Psychotherapy and Counselling. A Research Fellow and Professor (Life) of American Institutes. A member of working parties in Europe on systems and therapy.

During his varied career, Andreas held the positions of Overseas Marketing Executive, Production and Inventory Manager, Group Senior Systems Consultant, European Systems Manager, Principal Technical Adviser, Managing Director and Global Programme Manager with the multi-national organisations of International Computers Limited, Pitney Bowes, Plessey (GEC) Raychem, Engineering Industry Training Board (EITB), PsySys and EDS (Electronic Data Systems).

Participated as an information technology expert to European Union Commissions and as a therapeutic psychologist to the Institute of Psychology and Parapsychology and the Harley Street Centre, London. A published writer, poet and trainer, the subjects include Computing, Systems Methodology, Management and Psychotherapy. For achievements in Systems Engineering, Psychology and Directing, his biographical records are included in the directory of 'Who's Who In The World', published by Marquis of America and other international biographical publications.

Title:I.T. RISK MANAGEMENT IN SYSTEMS DEVELOPMENT

ISBN: 0 9527253 2 0

Author: Andreas Sofroniou
Copyright © Andreas Sofroniou 2002.
Published by: PsySys Limited,
33, Marlborough Road, Swindon SN3 1PH, U.K.
PRICE: U.K. £ 19-50

www.ingramcontent.com/pod-product-compliance
Lightning Source LLC
Chambersburg PA
CBHW051257050326
40689CB00007B/1225